The Discreet Charm of Protein Binding Sites

Joseph Yariv

The Discreet Charm
of Protein Binding Sites

 Springer

Joseph Yariv
Caesarea, Israel

ISBN 978-3-319-24994-0   ISBN 978-3-319-24996-4   (eBook)
DOI 10.1007/978-3-319-24996-4

Library of Congress Control Number: 2015957430

Springer Cham Heidelberg New York Dordrecht London

Printed on acid-free paper

Springer International Publishing AG Switzerland is part of Springer Science+Business Media (www.springer.com)

Preface

If I try to remember what brought me to writing a book about proteins, it was the mystery of why David Phillips and David Blow were not awarded the Nobel Prize. This was many years ago, but the mystery is still with us.

My interest in proteins was aroused by the proposition of Monod that the adaptive enzyme formation was controlled at the level of DNA by a repressor, a protein. While working as a postdoc in the USA, I found that Pauling's trivalent-antigen dyes could be used to isolate antibody to cytolipin H, a cerebroside in which a lactosyl residue is bound to the ceramide. My next challenge was to use such a dye with attached allolactose, a lac-operon inducer, to try and isolate the lac-repressor. But it was not to be! I remember well the jubilation I experienced when Müller-Hill and Gilbert isolated the lac-repressor and found that indeed it is a pure protein.

By the time of that discovery, A.J. Kalb (who later changed his name to A.J. Gilboa) and I working in Weizmann Institute of Science had isolated an L-fucose binding protein from a plant seed by such a trivalent dye with attached L-fucose. Unfortunately the isolated pure protein was a glycoprotein and thus was not considered suitable for attempting its crystallographic study. It was then that we turned to concanavalin A, a protein about whose structure nothing was known. It had been isolated many years earlier in a crystalline form by James B. Sumner, who received the Nobel Prize for demonstrating that an enzyme, the urease of Jack bean, is a pure protein. We determined the molecular weight of concanavalin A and its binding parameters for a glucoside and, most importantly, prepared some beautiful single crystals of it. All that remained to do was to undertake a crystallographic study of it. Since, at this time, there was no crystallographer in our institute, or in Israel for that matter, with experience in protein crystallography, my colleague was dispatched to the Laboratory of Molecular Biology in Cambridge to start such a study. On his return the work was continued with some members, visitors, and students in the unit of crystallography in the Department of Chemistry.

It was in Daresbury, this impressive structure where X-rays are produced at tremendous expense and resolved to beams of wavelength of a fraction of an angstrom, that I met John Helliwell. At that time the crystallographic study of concanavalin A had come to a standstill without solving the structure of its saccharide-binding site.

John was interested in concanavalin A because it gave relatively stable crystals. He agreed to work on the structure of its saccharide-binding site if I succeed to prepare well-diffracting crystals of its complex with a saccharide. Pretty soon I was on my way to York with a batch of well-diffracting crystals of the complex of concanavalin A with methyl-mannoside. John drafted Zygmunt Derewenda of York's Chemistry Department and their Xentronic/Nicolet area detector to collect the data. Soon Zygmunt sent a manuscript entitled "The structure of the saccharide-binding site of concanavalin A" to Max Perutz for submission to the *EMBO Journal*. Perutz asked for revisions! My preference was to avoid this delay by sending the paper to the *Journal of Molecular Biology* where our previous communication entitled "Properties of a New Crystal Form of the Complex of Concanavalin A with Methyl α-D-Glucopyranoside" had been published without much ado. But I was outvoted and the revised paper was accepted by the *EMBO Journal* two months later. This may sound petty to an outsider, but I have had my share of false priority claims to work that I pioneered in, and that is no trivial matter. It can poison the satisfaction one has from a well-deserved achievement.

This preface will not be complete without thanking the individuals who facilitated it. First and foremost I wish to thank Felix Frolow[1]. Felix is the guy who realized the solution of the structure of bacterioferritin of *E. coli* that I have obtained pure and crystallized. But that – as the saying goes – is another story. His help in realizing this book was on two levels. He helped with ideas of how to illustrate the binding of the various molecules to their binding sites and utilized his own data and the coordinates in the Protein Data Bank to produce the illustrations that enrich this book.

Without the guidance and the good cheer that I have received from the Springer editorial office, the experience of publishing a first book by an unknown writer would not have been the uplifting experience that I experienced. Having browsed in the library the chapter about proteins in the book *Soft Matter* by Roberto Piazza and liking what I read, I purchased the book. There on the back cover I found the endorsement of the book by Philip Ball. A recommendation by Philip Ball is good enough for me, and it propelled me to read the whole book. Having liked it, I wrote to the publishing editor, Maria Bellantone, and she was soon enthusiastic to be the editor of a book about proteins that I was writing. My correspondence with Maria was a very enjoyable experience. What is more, she makes you promise to do things that you never intended to do and to like it. She is of course aided, and protected (sic), by assistants. I had the good fortune to have Mieke van der Fluit and Annelies Kersbergen as intermediaries. Mieke possesses subtleness that succeeds to reassure you when in the blues and Annelies keeps you on a tight leash in a most endearing way.

Caesarea, Israel   Joseph Yariv

---

[1] Felix Frolow, a professor at Tel-Aviv University, died prematurely before this book was finished. He was mourned and is missed by his many colleagues and by the Israeli crystallographers of whom he was an outstanding member. He was a hands-on crystallographer with a good grasp on theory, having majored in physics. I have known Felix when he was the living soul of the Crystallography Laboratory of the Weizmann Institute since it started doing protein crystallography in the early 1970s.

Acknowledgments

First, I wish to thank all those authors whose work is listed in the references and made the writing of this book possible and, in particular, to those whose illustrations that are referred to have been a landmark in the maturation of Structural Biology. Unfortunately not all the listed illustrations could be reproduced because permissions to reproduce them could not be obtained. It is my hope that readers will look them up at their source. Then, I wish to thank the individuals who facilitated granting permissions I requested for their kindness and good cheer. They are Lucy Evans of IOP Publishing, Bristol, UK; Patricia Peticolas of Fundamental Photographs, New York, NY, USA: Elena Almoguera of Copyright Clearance Center, Danvers, MA, USA; Laura Stingelin of Elsevier, Philadelphia, PA, USA; Jane Ellis of University Science Books, South Orange, NJ, USA; and Stacey Riley of Nature Publishing Group & Palgrave Macmillan, London, UK. Many heartfelt thanks and appreciation are also extended to Haya Avital of the Design, Photography and Printing Branch, The Weizmann Institute of Science, Rehovot, Israel. Haya has adapted figures produced from our own and PDB – files by Felix Frolow and myself as well as some of the illustrations, both own and for which permissions were granted, to conform the request of the publisher.

Contents

List of Figures

About the Author

Joseph Yariv graduated from the Hebrew University in Jerusalem with a Ph.D. in biochemistry. After postdoctoral studies at the Sloan Kettering Institute and Columbia University in New York, USA, he joined the department of biophysics of the Weizmann Institute of Science in Rehovot, Israel, where he worked until his retirement in the position of senior scientist. His work dealt with protein isolation, crystallization, and structure solution. He was the first to label a methionine in the active site of β-galactosidase of *E. coli*. He produced crystals of concanavalin A complexes with methyl-glucoside and with methyl-mannoside and participated in solving the structure of this protein binding site for saccharides. He collaborated with physicists at the Hebrew University in Jerusalem in studying by Mossbauer spectroscopy the state of iron in *E. coli* that led to the isolation of bacterioferritin, the first ferritin-like molecule to be found in bacteria and named as such, and solution of its structure.

Chapter 1
Introduction

What Is a Protein?

The study of proteins in the last century has provided insights into the functioning of living matter that have hardly a counterpart in the study of other living matter constituents such as nucleic acids, carbohydrates or lipids. Proteins have been studied for a long time without it being realized that they are molecules. Nowadays, what protein molecules look like is displayed in museums (Paterlini 2008).

Much of our understanding of protein molecules coincided with the understanding of how proteins are synthesized. This did not occur until 1960s when it was established that a nucleic acid molecule, the so-called messenger RNA, is the template that determines what kind of protein is being synthesized and specifies the exact size/weight of such a molecule. Messenger RNA is a linear stretch of nucleotides that specifies the linear stretch of the amino-acids of the protein molecule including its start and finish. The instruction that the messenger RNA gives to the protein synthesizing machinery of the cell is the sequence of the nucleotides in the chain that is read in units of three, the sequence of the three nucleotides specifying the amino acid that will be synthesized. The nucleotide can be any of the four nucleotides carrying the following bases: adenine (A), guanine (G), uracil (U) and cytosine (C). The sequence of three of these four bases, the triplets, specify the twenty amino-acids that can be used in the synthesis of the protein chain as depicted artistically on page 22 of a book entitled "the Structure and Action of Proteins" by Dickerson and Geis 1969. Of the 20 α-amino-acids only one is optically inactive, the glycine. It is composed of just two linked carbon atoms, one a carboxylic acid and the adjacent carbon atom, the α-carbon atom, substituted with an amine residue and two hydrogens. All the others have an organic grouping on the α-carbon atom that must be of the L-configuration to be utilized in the synthesis of the protein. The smallest of these, the alanine, has just a methyl group on the α-carbon atom. The others have groupings of different size, shape and properties and are listed in Table 1.1. In a protein molecule successive amino-acids are linked by a peptide

© Springer International Publishing Switzerland 2016
J. Yariv, *The Discreet Charm of Protein Binding Sites*,
DOI 10.1007/978-3-319-24996-4_1

Table 1.1 A List of the 20 α-Amino-acids that Proteins are Composed-of (Arranged according to their rising weight)

Trivial name	Chemical formula	Abbreviation	Single letter code
Glycine	$C_2H_5NO_2$	Gly	G
Alanine	$C_3H_7NO_2$	Ala	A
Serine	$C_3H_7NO_3$	Ser	S
Proline	$C_5H_8NO_2$	Pro	P
Valine	$C_5H_{10}NO_2$	Val	V
Threonine	$C_4H_9NO_2$	Thr	T
Cysteine	$C_3H_7NO_2S$	Cys	C
Leucine	$C_6H_{13}NO_2$	Leu	L
Isoleucine	$C_6H_{13}NO_2$	Ile	I
Asparagine	$C_4H_8N_2O_3$	Asn	N
Aspartic acid	$C_4H_7NO_4$	Asp	D
Lysine	$C_6H_{14}N_2O_2$	Lys	K
Glutamine	$C_5H_{10}N_2O_3$	Gln	Q
Glutamic acid	$C_5H_9NO_4$	Glu	E
Methionine	$C_5H_{11}NO_2S$	Met	M
Histidine	$C_6H_9N_3O_2$	His	H
Phenylalanine	$C_6H_{11}NO_2$	Phe	F
Arginine	$C_6H_{14}N_4O_2$	Arg	R
Tyrosine	$C_9H_{11}NO_3$	Tyr	Y
Tryptophan	$C_{11}H_{14}N_2O_2$	Trp	W

bond, a bond that is formed between the carboxyl of one amino-acid and the amine of the amino-acid that succeeds it with extrusion of a molecule of water. Synthesis of a protein specified by messenger RNA proceeds from the carboxyl of the first amino-acid whose amine residue is left intact to the last whose carboxyl residue is left intact and the residues are numbered accordingly.

The Heroes!

Already in the early 1920s J B Sumner provided unequivocal evidence that a protein is a chemically pure entity when he crystallized the enzyme urease that he isolated from the seeds of Jack bean and which catalyses the splitting of urea into ammonia and carbon dioxide. At that time, the established opinion did not credit proteins with catalytic property and considered them rather as ill-defined macromolecular carriers of various chemicals with biological properties. Sumner was awarded the Nobel Prize in 1946 (Sumner 1946), some 20 years after his revolutionary breakthrough, and he shared it with J H Northrop who has by then crystallized a number of enzymes. Since then the number of Nobel Prize Laureates in Chemistry recognized for their research of proteins in the last century amounts to more than 20. To name but a few: F Sanger for determining the primary structure of insulin to wit, the sequence of amino acids in the two chains that compose an insulin molecule; J C Kendrew and M F Perutz for solving the three-dimensional (3-D) structures of myoglobin and haemoglobin, the first proteins whose 3D-structure has been solved by X-ray crystallography. Equally significant, even if not recognized by award of Nobel Prizes, was the solution of the structures and of the catalytic mechanisms of two enzymes: of lysozyme by the team headed by D C Phillips (Blake et al. 1965) and of chymotrypsin by D M Blow and his collaborators (Matthews et al. 1967; Blow et al. 1969). Another individual whose contribution at that time has not been recognized by the award of a Nobel Prize is F M Richards, famous for his work on the S-peptide of ribonuclease that demonstrated for the first time the importance of shape and van-der-Waals interactions in the assembly of the polypeptide chain into the 3-dimensional entity that is the energetically favored and functionally competent protein molecule (Richards 1958, 1997).

Institutionalizing Protein Research

The study of protein structure necessitated cooperation of scientists from different fields and the provision of sophisticated equipment. In mid-twenty century this was to be found only in a few centers of excellence, even if at that time scientists were too discreet to claim this distinction for themselves or their laboratories. Until this time X-ray data collection was done on home-made equipment utilizing a cathode tube as generator of X-rays. Soon these were supplemented by synchrotrons where X-rays are generated by electrons accelerated to speeds approaching that of light and at atomic reactors that generate neutrons. (Is it but a coincidence that the English synchrotron was in Daresbury, the birthplace of Lewis Carroll, the author of "Alice in Wonderland"?) Thus within the space of a few years protein structure research became big-science. Factory-size laboratories are used to provide beams that probe the structure of single crystals of pure proteins and of complexes of proteins with other molecules.

Proteins and Biological Sciences

The impact that the study of protein structure has had on biological sciences cannot be overestimated but its philosophical impact has been greatest on the theory of evolution. Its impact on the way we think about the past has been outstanding and its mechanisms continue to challenge the biologist. Ironically, genetics, in the form of the DNA sequences of the relevant genes, revealed the close relationship of the organisms that compose our biosphere. To say that there is a 98 % correspondence between the genes of a chimpanzee and that of a human can be shocking to some but it is a scientifically established fact. But what is even more astounding is the fact that genes of organisms as distant from man as a fly or a plant show homology which in everyday language means that stretches of nucleotide sequences in the respective genes are similar and that the proteins which they produce perform similar functions in these distant organisms. These results are very solid scientifically and science progresses only by explaining facts for which there is no valid explanation. The similarity of the genes of plants, flies and humans calls for a scientific explanation as does the similarity in the genetic make-up between a chimpanzee and a human being.

Proteins are both the components and the engines of our bodies. It is estimated that human body is composed of some 30,000 kinds of proteins. Thanks to genetics we know how the sequence of nucleotides in a gene specifying a certain protein is translated to produce a sequence of amino-acids of a given size that is the protein molecule. However, the properties of the protein, even if specified by this sequence, are not the sequence itself but the structure in three dimensions (3-D structure) that it acquires on interaction with water and often on interaction with other molecules. A typical protein molecule can be compared to a string of beads collapsed onto itself like a heap, as pictured originally on Plate 9 in a book entitled "The Thread of Life"(Kendrew 1966); except that the beads in this heap each has an identity, being one of the twenty odd α-amino acids. Furthermore, they are all of L configuration, their optical isomers, D α-amino acids, being excluded. In anticipation of the anthropomorphic principle of the cosmologists, Linus Pauling has speculated that the reason that proteins are composed solely of L-amino acids could be that were they to be composed of the mixture of D- and L-amino acids they could not form the secondary structure of the a-helix, the form that he predicted to exist in proteins (Pauling 1993; Pauling et al. 1951). This prediction was based on the results obtained from X-ray diffraction study of crystals of peptides that showed planarity of the peptide bond (Corey 1938) and is illustrated on page 13 of Dickerson and Geis (1969). The comparison of a protein molecule to a collapsed string of beads is true of almost all the different kinds of protein molecules. Many are water soluble but many are components of supra-molecular structures. Water soluble proteins were the ones that were studied, especially if they could be isolated pure of contaminants and if single crystals of them could be produced. What we know about proteins as molecules comes primarily from X-ray crystallographic study of protein single crystals (Fig. 1.1). These studies combined with the study of proteins in solution and with theoretical considerations form the basis of our understanding of what proteins are and of the interactions they make with other molecules.

Protein Structure Solution

The history of crystallography is an epic. It grew out of the discovery by Max von Laue that crystals of pure chemicals diffract X-rays and the subsequent utilization of X-ray diffraction by the father-and-son team, W Laurence Bragg and William H

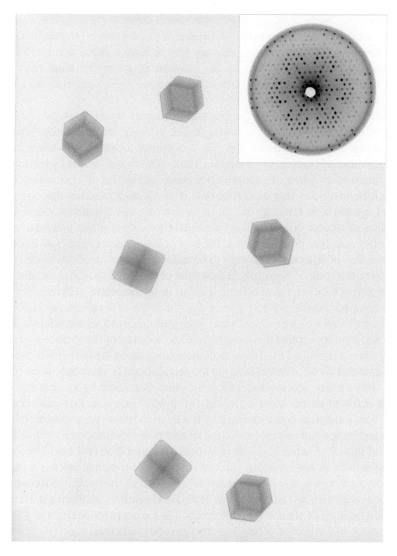

Fig. 1.1 A photograph of some uniquely attractive crystals of bacterioferritin and a precession photograph down the crystal pseudo threefold axis (Smith et al. 1989; Frolow et al. 1993)

Bragg, in solving the structures of minerals. This was then extended to solution of the structures of organic molecules and eventually of proteins. Its impact on organic chemistry was poetically expressed by R.W. James in a book that has seen many editions (James 1953). Thus he writes "………. In many crystals, however, and particularly in the crystals of organic compounds, there is a well-defined molecule. The atoms in this molecule are linked together not by ionic or electrostatic forces, but by some form of covalent binding………and the whole crystal is built up of such molecules, held together by the residual forces of attraction between

them…………. Such crystals are, as a rule, much softer, far less rigid in structure
…… One of the interesting results of the analysis of organic crystals is that structural formulae of the organic chemists are seen to have a physical existence. The
benzene ring is a ring; naphthalene has a double ring; and the long-chain compounds are really long chains of carbon atoms whose length and configuration can
be determined."

While nowadays solution of the structure of a small organic molecule by X-ray
diffraction is done routinely by technical staff or by dedicated commercial laboratories, this was not so at the beginnings. The first breakthrough was the determination
of the structure of oxalic acid (Robertson 1938). This was followed by the solution
of the structures of a host of aromatic compounds and by that of saccharides. While
these methods were at first successful only if the number of atoms that composed
these molecules were few, they did not suffer from another limitation to the solution
of protein structures: crystals of organic small-molecules diffract to a high resolution while crystals of proteins do not. Because of this property there is no limitation
to the number of reflections that need to be collected from a small-molecule crystal
to perform the computations that improve the accuracy of bond-lengths distances
and the angles between the neighboring atoms in the examined structure. This particular computational procedure goes under the name of least squares refinement
and in small molecules produces bond-length distances to within 0.01 Å. When
Corey and Pauling applied these methods to the solution of the structures of small
peptides they established the distances of the atoms making-up the peptide bond and
discovered that all the atoms making-up the peptide bond lie in a single plane (Corey
1938). This finding was instrumental in the prediction made by Pauling that polypeptide chains might pack as α-helices of a given pitch in protein molecules (Pauling
1993). This prediction was confirmed by Perutz even before the structures of myoglobin and haemoglobin were solved, as he told in his reminiscences of this heady
period of protein structure research (Perutz 1987). Validation of the existence of the
Pauling α-helix in haemoglobin validated as well the assumptions that went into
constructing it to wit, that the imino hydrogen atoms of the backbone form hydrogen bonds with carbonyl oxygen atoms of the backbone thus stabilizing it. Here was
one solid example of what is the conformation of the polypeptide chain in a protein
molecule or, in the specialist language, its secondary structure.

The principle of structure solution by means of X-ray diffraction is based on the
fact that electrons scatter X-rays more efficiently than protons. The electron scattering power of atoms increases with the number of electrons they contain thus allow-
ing recognition of atoms in molecules composed of many different atoms. This is a
general property of atoms, applying as well to molecules in the gaseous state as in
solids. The result of a satisfactory X-ray diffraction experiment is the electron density imprint of the molecule (Fig. 1.2). When the chemistry of the molecule is
known, it is possible to fit the respective atoms of the molecule into the electron
density projection or its map. X-ray diffraction of crystals is used primarily to determine the electron density of molecules of known chemistry but unknown structure.
When a molecule was chosen whose shape and size were estimated from chemical
evidence and optical and magnetic data, the structure as determined by X-ray

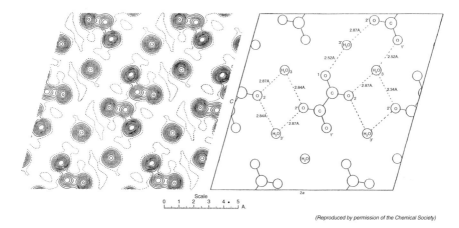

Fig. 1.2 A projection of the electron density of oxalic acid with an accompanying projection of atom distances (Robertson 1938, p. 358)

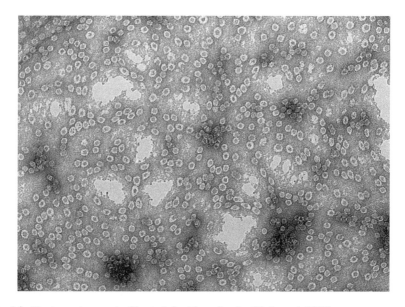

Fig. 1.3 Electron micrograph of bacterioferritin molecules (Yariv et al. 1981)

diffraction was found to be in very good agreement with the theoretical expectations (Abrahams et al. 1949).

The difficulty we have in grasping a structure of a molecule, and especially of molecules as big as proteins, is that molecules occupy space. A protein molecule as seen in an electron micrograph gives its shape but no information about its structure (Fig. 1.3). (This statement which was true until quite recently is contradicted in a recent article by Liao et al. (2013), and reviewed by Henderson (2013) in the same issue of Nature, where the structure of heat-sensing ion-channel has been solved to a resolution of 3.4 Å by electron microscopy and to quote Henderson "good enough for amino-acid side chains and β-sheets to be recognized, and the polypeptide backbone of the protein to be traced".) X-ray diffraction, on the other hand, gives the position of practically each and every atom of the structure if data are collected to a high enough resolution. The first protein structure to be solved by X-ray diffraction was that of myoglobin and this was done first to 6 Å and then to 2 Å resolution. Even at 6 Å resolution the amino-acid sequence of the protein could be fitted into the experimental electron density map as could the non-protein chromophore of this molecule, the haem. The outstanding result of this work, beautifully popularized by Kendrew, was that it demonstrated that most of the protein chain in this molecule assumed the geometrical form of α-helix, as predicted by Linus Pauling (Kendrew 1961). Soon, another geometrical form of the polypeptide chain in a protein molecule, of β-sheet, was observed in the solved structure of the enzyme lysozyme (Blake et al. 1965). Models of protein structures at this time utilized the available solid models of organic chemistry such as the 'ball and stick' model or space filling models. 'Ball and stick' model used small rods representing bond distances between the atoms to connect small balls representing atoms. Space filling models of the atoms, also called CPK (Casey, Pauling, Kaltun) models, represented the van der Waals radii of the atoms and were connected by short links inserted into receptacles on the surface of the atoms to present a continuous surface of the resulting molecule. Still another model, the Dreiding model, addressed the possibility of conformational change in the built structure. All these aids to visualizing the molecular structures were soon replaced by computerized graphics that allows viewing the structures in three dimensions.

When a crystal is irradiated with X-rays, the diffracted rays contain information not only of the structure of the molecules composing it but also of the disposition of these molecules. Crystals do come in different geometrical forms and the rules that govern the disposition of the objects composing a crystal have been worked out by mathematicians starting with L. Euler in the eighteenth century. It is thus a well established fact that crystals can have no symmetry or a twofold, threefold, fourfold and sixfold symmetry but no fivefold symmetry or a symmetry higher than sixfold. Symmetry considerations permit 7 crystal systems: triclinic, monoclinic, orthorhombic, tetragonal, trigonal, hexagonal and cubic. Depending on whether the relationship between the objects composing the crystal is a simple rotation, reflection or a combination of these operations with a displacement, a crystal can belong to any of mathematically possible 230 space groups. Because proteins are chiral objects, composed of L-amino acids, reflections are impossible for protein crystals and a protein crystal can belong to one of only 65 space groups.

X-rays diffracted by a crystal are recorded by an electronic counter or on film that record diffraction maxima of the diffracted rays. These maxima, or Bragg reflections, produce also a lattice, the so-called reciprocal lattice, which bears a geometrical relationship to the crystal lattice. As can be seen in Fig. 1.1, reciprocal lattice is pretty. It is always of a higher symmetry than the symmetry of the crystal that produces it. Still the reciprocal lattice is directly related to the real lattice of the crystal and allows extraction of important information from the disposition of the diffraction maxima such as its symmetry and the dimensions of the unit cell of the crystal. The unit cell of the crystal is the object whose symmetry operations in a given space group generate the crystal lattice. Knowledge of the space group of a crystal facilitates the strategy of data collection to solve the structure of the asymmetric unit of the crystal, the smallest structural unit of the crystal. A unit-cell of the crystal can have just one asymmetric unit in the case of the triclinic space group and up to many asymmetric units related by the symmetry operations of the space group. This object cannot be smaller than a protein molecule if it is composed but of a single subunit. The quantity of data necessary to solve the structure of the asymmetric unit will depend on its size and on the resolution to which the structure will be solved. The dependence of resolution on the number of reflections that it is necessary to collect to solve a given structure to a given resolution is very steep. As an example, the numbers given by Blow in his recent book are quoted (Blow 2005): "if all that is needed to solve a protein structure of 20 000 daltons to a resolution of 4.5 Å are 1080 reflections, a solution to a resolution of 2.0 Å requires 12,000 reflections and one to 1.34 Å requires 41,580 reflections".

The recorded diffraction intensity maxima are used to derive the electron density map of the molecule composing the crystal by combining it with information about the phase of the diffracted X-rays. Difficulty in determining the phases of the diffracted rays were the main bottleneck in solving protein structures. This bottleneck was surmounted by a technique that goes under the name 'isomorphous replacement'. In this technique a heavy atom, such as mercury (Hg) for example, is incorporated into the protein molecule and if thus modified protein produces a crystal of the same unit-cell dimensions and of the same space-group as the unmodified one namely, an isomorphous crystal, the differences in the intensities between this isomorphous pair are used. Except that the differences of one isomorphous crystal with a heavy atom in it are not enough. Usually crystals of three different heavy atom derivatives, or rather isomorphous crystals of such derivatives, are needed to calculate the phases and this method therefore goes under the name 'multiple isomorphous replacement' or, in short, MIR. Extraction of information about the phases of the recorded reflections abounds in methods known by their anagrams. The most expressive of these is MAD for 'multiple wavelength anomalous dispersion'. Its advantage is that only one crystal is needed if the protein incorporates an anomalous scatterer (Karle 1989). I shall not attempt to explain the physical basis of this method except to say that, again, it is produced by heavy atoms, the anomalous scatterers. A very popular scatterer for MAD has become selenium (Sn) that can be incorporated into proteins as the unnatural amino acid selenomethionine where it replaces the natural amino acid methionine with which it shares many chemical properties (Hendrickson et al. 1990). Then there are methods that are a combination of the two aforementioned methods such as, SIRAS for 'single isomorphous replacement with anomalous scattering' and MIRAS for 'multiple isomorphous replacement with anomalous scattering'. A more recent addition to the armoury of tools used to solve the phases of the reflections from protein crystals is collecting data from protein crystals exposed to xenon (Schiltz et al. 2003). I shall tell about xenon binding later, when I introduce 'protein binding-sites'.

Once the phase of the diffracted waves is known or guessed (sic), electron density of the molecule is calculated by means of a mathematical procedure, the Fourier synthesis, developed in the eighteenth century by the mathematician Joseph Fourier in connection with his studies of heat conduction in solid bodies (Bracewell 1989).

There are not many protein structures that have been solved to a resolution higher than about 2 Å and many to as low a resolution as 3–4 Å. The fact that reasonable structures have been extracted with such limited electron density data is due to procedures that go under a collective name of refinement. However, refinement of a protein molecule is nothing like the least square refinement used to refine small-molecules, as was mentioned above. The first stage in interpreting the electron density of a protein molecule is fitting its amino-acid sequence into the electron density map. This stage is followed by a series of computational procedures designed to correct deviations of the positions of the atoms within the amino-acid chain from the theoretically correct, or permitted, ones. These operations, the refinement, when successful in improving the correctness of the protein structure are expressed by a statistical factor, the reliability or R factor, which includes both experimental and calculated terms, similarly to the one used in least square refinement of small molecules. Successful refinements result in the R factor decreasing. But, whereas small molecule refinements result in R factors of 0.05, most well resolved protein structures give R factors in the range of 0.2–0.3. Refinement of protein structures is done in stages and usually starts with low resolution data that depend to a large degree on the quality of the experimentally determined phases. The resolution to which the intensity data are collected, their quality and their completeness will determine the quality of the refined structure. It should be borne in mind that structure determination is an exact science and that the expected quantity of data from a known number of atoms in a given space group can be calculated for any given resolution. Experimental determination of a molecule structure also involves calculations based on the number of the atoms. A position of an atom can be defined by four parameters, the three coordinates of space and an isotropic temperature factor that measures the vibration of the atom relative to its equilibrium position. A rule of thumb suggests that for the sake of reliability there should be an excess of three observations for each atom parameter and another rule of thumb suggests that only reflections with small standard error be utilized for the computations. As a matter of fact, redundancy of data is utilized whenever possible. Refinement procedures for proteins introduce a number of requirements, or limitations, on the resulting structure such as bond lengths and bond angles between atoms which have to conform to values found for small peptides solved at very high resolution. The ones that are concerned with the angles of the α–carbon, the phi (φ) and psi (ψ) angles (as illustrated on page 25 of Dickerson and Geis 1969), are listed in a plot, the Ramachandran plot, that accompanies every solved protein structure on its inclusion in a protein data-bank. For a glimpse of this outstanding man and the story behind this famous plot see Ramachandran obituary published in Nature (Vijayan 2001).

In the above description of what principles and methods are utilized in solving the structure of a protein molecule I chose to omit sophisticated and mathematically heavy approaches that were utilized to solve the structures of small molecules from diffraction intensities without knowledge of the phases of the scattered waves. I also omitted to mention the different computer programs that are used to facilitate the calculation of the enormous quantity of numerical data that need to be processed in, for example, the refinement or in producing the electron density of a molecule. No justice however can be done to the Molecular Replacement method of extracting the missing phases from the diffraction data of a crystal that contains molecules similar to molecules whose structure was solved already without mentioning that it depends on mathematical treatment and an extensive utilization of computer programs. This is a method that has been called "A method ahead of its time" in Milestone 13 of the Milestones timeline: Nature Milestones in Crystallography that is available free by Nature Online (Doerr 2014).

And this is what David Blow had to say about this method, that he initiated with Michael Rossmann (Rossmann and Blow 1962), in a book entitled "Outline of Crystallography for Biologists" (Blow 2005). He divides the problem of solving the unknown structure of a molecule in a crystal X by using the information available about the structure of a similar molecule M in a different crystal into four steps. First step is to find the angular orientation of the M-like molecule in the crystal of X. Next, given this angular orientation, to find what point in X corresponds to a chosen origin position for M. Then one generates a hypothetical molecular structure of the unknown crystal X that is composed on M-like molecules. Finally, one improves this hypothetical structure by using the diffraction measurements of X. In the context of Protein Binding-sites, Blow's chapter on Molecular Replacement includes a section entitled "Application of molecular replacement to complexes like enzyme-substrate or enzyme-inhibitor complexes".

A complete departure from the need of X-ray diffraction of protein crystals to determine the structure of proteins is embodied in a discipline that goes under the name of Molecular Modeling. Molecular modeling attempts to solve protein structures by the fundamental rules of physics, both Newtonian and quantum-mechanical. It bases its attempts on the knowledge of the primary structure of the protein namely the knowledge of the size of the polypeptide chain and the sequence of the amino-acids that compose it. Considering that knowledge of the primary sequence of a protein can be extracted from the sequence of the nucleotides that compose its gene, molecular modeling attempts to determine the 3-D structure of a protein even before it has been isolated. Thus molecular modeling attempts to determine the molecule's secondary structure, its tertiary structure and the quaternary structure of protein molecules composed of subunits, all this in-Silico. It certainly attempts and even succeeds to solve the interaction of small molecules with proteins and this is its appeal to pharmaceutical companies that use molecular modeling to test innumerable molecules as candidates for development of drugs that interact with proteins of choice known to be involved in a malady. A critical review of Molecular Modeling achievements in drug design and its future is contained in an article by Derek Löwe entitled "Molecular modeling's $10-million comeback?" (Löwe 2010). Molecular

Modeling evolved from Molecular Graphics and a pioneer of both these methods was none other than Cyrus Levinthal another physicist that has been won-over to biology and excelled in it (Magagno et al. 2014).

The story of protein structure solution will not be complete without mentioning the artistry that went into picturing the emerging structures. If a confirmation is necessary that it is art, this can be supplied by the following anecdote dating from the late nineteen-forties whose authenticity I cannot ascertain. Pablo Piccaso was visiting Desmond Bernal in his laboratory in Birckbeck College, London. On being shown by Bernal into a room where an immense structure of a ball-and-stick model of a protein molecule was displayed, Picasso raised his hands and exclaimed "what a splendid imagination". But the credit for employing art in the service of picturing protein structures that have not been surpassed until this day must go to Irving Geis (Kemp 1998). This came to its full fruition when Geis illustrated the first book on protein structure written by R E Dickerson (Dickerson and Geis 1969). Since then attempts have been made to picture the amino-acid sequence of a protein molecule as a ribbon and to assign names to its twists namely, to the elements of its secondary structure. Thus a nomenclature for a protein molecule anatomy was devised utilizing such trivial names as "jelly roll", "barrel" etc. (Swindells et al. 1998). Solution to picturing the structure of proteins was achieved when films of protein structures became available, however these can hardly be of use on the printed page. In these pages I shall illustrate surface views of limited regions of proteins, the binding-sites, to which space-filling models of small molecules are bound.

Where Is It Headed?

The challenge that proteins continue to present to scientists is of two kinds. One is the ability to know the 3-D structure of a protein from its so-called 'primary structure'. Despite all the accumulated knowledge about the 3-D structure of the known proteins, it is not possible to know the 3-D structure of a protein unless a crystal of it is produced and its structure solved. The 'primary structure' of a protein, being the sequence of its building blocks, or the sequence of the beads in the string in the analogy above, can be determined directly, once a sufficiently pure protein is available, or it can be divined from the structure of its template, the messenger RNA, or from the structure of the gene that specifies this template. The challenge of knowing the 3-D structure of a protein from its 'primary structure' is both practical and fundamental. The fundamental difficulty is inherent to the protein structure. Because even a small protein molecule consists of a few thousand atoms and each atom has many degrees of freedom - some more, some less, depending on their position in the structure – assigning a unique position to each and every atom in the structure has until now defeated the attempts of the theoreticians. The practical difficulty is that without this capacity the 3-D structure of most human proteins cannot be established, be it because most proteins are not available in quantity sufficient for X-ray crystallographic analysis or because analysis of tens-of-thousands proteins is not a

practical proposition even if each and every protein in the human body were produced in quantity, purity and quality of crystal to make such an analysis feasible.

Another challenge is to explain the known properties of living organisms in
terms of the proteins that compose it. Some of these properties have already been
explained. The first startling success was that of explaining the metabolic cycles,
those of fermentation and respiration, photosynthesis and the bio-synthesis of
amino-acids and nucleotides by the respective catalytic proteins, the enzymes. This
was followed by explanation of the inheritance of the genetic content and its expression by a host of specific enzymes and regulatory proteins, culminating in the mechanism of the bio-synthesis of the proteins themselves. The conceptual leap that is
now being called for is to explain in terms of proteins the concerted activity of a
living organism, the activity that underwrites the organism's individuality.

An eucaryotic organism, even a unicellular one, comes equipped with two kinds
of proteins that are a prerequisite for maintaining the individuality of the organism.
One is the so-called tubulin, a protein that composes the cytoskeleton, a net of
water-filled tubules that maintain the shape of the cell. Tubulin is also to be found in
the cilia, protrusions from the body of the cell that in unicellular organisms, the cili-
ata, propel the organism in its search for food or mate. And most prominently, tubulin composes the so called spindle that aligns and separates the two sets of
chromosomes before cell division (Fig. 1.4). The other kind of protein encompasses
a large variety of proteins that reside in the membranes of the cell. These are first
and foremost the cellular membrane but also the nuclear membrane and the membrane of the mitochondrion, the metabolic powerhouse of the cell. Analysis of the
total genome of a yeast suggested that 40 % of its proteins are membrane proteins
(Goffeau et al. 1993).

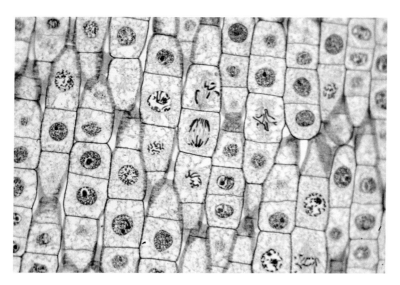

Fig. 1.4 The mitotic spindle. Mitotic Division in onion root tip (By permission of Fundamental
Photographs)

In multi-cellular organisms, again, and in particular in the higher organisms like the mammals, specialized proteins make their appearance. Thus the most abundant protein in the human body, the collagen, is present in the intercellular spaces where it connects the various organs. In the muscle, myosin and actin are prominent and instrumental in muscle contraction. Antibodies are produced in specialized cell and are involved in recognition of foreign bodies introduced into the organism, whether bacteria or viruses introduced by infection or transplants of foreign tissue. Protein hormones are produced in special glands and affect distant organs.

All the advances in the understanding of proteins are explained by physics and chemistry and did not necessitate postulates outside these disciplines. If anything, explanation of the phenomena encountered in the study of proteins is based on clas- sical physics without it having been necessary in the past to invoke the more abstract quantum mechanics. However, it should be remembered that molecules are objects that do exhibit quantum-mechanical phenomena (Ball 2003; Hackermüller et al. 2003) and that this must be true for protein molecules as well. Already in the 1940s, F. London suggested that macromolecules may exhibit long-range van der Waals interactions not observed in small molecules (London 1943). R Penrose, recently, is advancing an even more revolutionary notion namely, that human consciousness is a quantum mechanical macroscopic phenomenon associated with the universally present cytoskeleton made-up of a protein, the tubulin (Penrose 1994). In this he is heir to the famous scientists in the past who did not shy biological objects if phe- nomena could be observed in them hitherto not observed in non-animate matter.

Protein Binding-Sites

The above summary of protein research beginnings, achievements and aims omits the tedious procedures involved in the isolation and identification of the various proteins. While the actual methods of isolating proteins from the tissues, organs or packed unicellular organism are hardly illuminating even when presented in labora- tory manuals, the thinking that went into it explains the mental processes that guided this research from its very beginning. I have selected to tell about the concept of a protein binding-site because it derives from the anthropomorphic concept of recog- nition, and thus is easy to comprehend, and because it was fruitful when applied to protein isolation. It is no exaggeration to say that of all macromolecules known to man, whether of biological or synthetic origin, no molecule exhibits the kind of selectivity exhibited by proteins in the binding of small molecules. Furthermore, only in proteins does one observe the so-called 'site-site interaction' that informs a molecular assembly whether its sites are occupied or empty and that, sometimes, has far-reaching metabolic or physiological consequences.

Binding-sites are a precondition for the activity of enzymes where they are called 'active-sites'. Binding of a substrate by an active-site of an enzyme is followed by a series of chemical reactions which result in products that are characteristic of the enzyme. For example, the Jack bean urease already mentioned in this introduction,

catalyses the splitting of urea into ammonia and water. Competitive enzyme inhibitors, on the other hand, bind to the active site of the enzyme without being split.
Binding of small molecules by proteins is, however, not restricted to enzymes and
is the topic of this book. Saying that a protein recognizes a molecule that binds to it
does not say much about the physical processes involved in this binding, while
qualifying this kind of binding as reversible suggests that the forces involved are
weak. A summary of intermolecular forces is given in Table 1.1 of J D Bernal in his
introduction to a discussion entitled 'Structure Arrangements of Macromolecules'
(Bernal 1958). Van der Waals forces and hydrogen-bonds are the main forces oper-
ating in the interaction of protein molecules with other molecules if one discounts
the repulsive forces that operate between atoms and molecules. A very good introduction to the binding of small molecules to protein is the binding of atoms of the
noble gas xenon. The very first observation of an atom of xenon in a protein was in
myoglobin (Schoeneborn et al. 1965). This work has shown that while the atoms of
a protein are closely packed there exist empty spaces in this structure. A very easy
to understand explanation of xenon binding to proteins is to be found in a section
entitled 'Molecular Forces in Xenon/Krypton-Protein Interactions' to be found on
pages 89–94 of a publication by Schiltz, Fourme and Prangé entitled 'Use of Noble
Gases Xenon and Krypton as Heavy atoms in Protein Structure Determination' that
was published in 2003 in the 374th volume of Methods in Enzymology (Schiltz
et al. 2003). Here I reproduce the first paragraph of this section that is self-
explanatory. "As noble gas atoms have a zero charge and are spherically symmetric,
Coulomb interactions, hydrogen bonding and dipole-dipole interactions cannot be
involved in the binding of xenon and krypton to proteins. Thus the only possible
attractive interactions between noble gas atoms and proteins are charge-induced,
dipole-induced, and London (dispersion) forces. The key physical parameter in
these interactions is the electronic polarizability of the noble gas atoms. The usual
repulsive forces between atoms and molecules that are in close contact with each
other also play an important role because they determine the minimum size that the
cavity must have for xenon or krypton to bind to it." Incidentally, these authors are
the ones who have conceived and realized this new method of protein structure solution. The general reader should not be discouraged of having to seek out this refer-
ence in specialized libraries, since Methods in Enzymology can now be accessed on
the internet.

References

Abrahams SC, Robertson JM, White JG (1949) The crystal and molecular structure of naphtha-
    lene. II. Structure investigation by the triple Fourier series method. Acta Crystallogr
    2:238–244
Ball P (2003) Molecules of life come in waves. Nature (Science Update, Sept 6)
Bernal JD (1958) Structure arrangements of macromolecules. Discuss Faraday Soc 25:7–18

Blake CC, Koenig DF, Mair GA, North AC, Phillips DC, Sarma VR (1965) Structure of hen egg-white lysozyme. A three-dimensional Fourier synthesis at 2 Å resolution. Nature 206:757–761

Blow D (2005) Outline of crystallography for biologists. Oxford University Press, New York, p 214

Blow DM, Birktoft JJ, Hartley BS (1969) Role of a buried acid group in the mechanism of action of chymotrypsin. Nature 221:337–340

Bracewell RN (1989) The Fourier Transform. Sci Am:62–69

Corey RB (1938) The crystal structure of diketopiperazine. Proc Natl Acad Sci U S A 60: 1598–1604

Dickerson RE, Geis I (1969) The structure and action of proteins. Harper & Row, Publishers, New York

Doerr A (2014) A method ahead of its time. Nature, Milestone 13/Nature Milestones/Crystallography, August 2014

Frolow F, Kalb AJ, Yariv J (1993) Location of haem in bacterioferritin of E. coli. Acta Crystallogr D 49:597–600

Goffeau A, Nakai K, Slonimski P, Risler J-P (1993) The membrane proteins encoded by yeast chromosome III genes. FEBS Lett 325:112–117

Hackermüller L, Uttenthaler S, Hornberger K, Reiger E, Brezger B, Zeilinger A, Arndt M (2003) Wave nature of biomolecules and florofullerenes. Phys Rev Lett 91:090408 (1–4)

Henderson R (2013) Ion channel seen by electron microscopy. Nature 504:93–94

Hendrickson WA, Horton JR, LeMaster DM (1990) Selenomethionyl proteins produced for analysis by multiwavelength anomalous diffraction (MAD): a vehicle for direct determination of three-dimensional structure. EMBO J 9:1665–1672

James RW (1953) X-Ray crystallography, 5th edn. Methuen & Co. Ltd, London, p 93

Karle J (1989) Macromolecular structure from anomalous dispersion. Phys Today 42:22–29

Kemp M (1998) Kendrew constructs; Geis gazes. Nature 396:525

Kendrew JC (1961) The three-dimensional structure of a protein molecule. Sci Am 205:96–110

Kendrew JC (1966) The thread of life: an introduction to molecular biology, figure 9. G. Bell and Sons Ltd, London

Liao M, Cao E, Julius D, Cheng Y (2013) Structure of the TRPV1 ion channel determined by electron cryo-microscopy. Nature 504:107–118

London F (1943) Intermolecular attraction between macromolecules. Surf Chem 21:141–149

Löwe D (2010) Molecular modellings $10 million come back? Nature 7 May, pp 1–3 (online)

Magagno E, Honig B, Chasin L (2014) Cyrus Levinthal 1922–1990. Biographical Memoirs, National Academy of Sciences

Matthews BW, Sigler PB, Henderson R, Blow DM (1967) Three-dimensional structure of tosyl-a-chymotrypsin. Nature 214:652–656

Paterlini M (2008) A protein ghost etched in glass. Nature 452:155

Pauling L (1993) How my interest in proteins developed. Protein Sci 2:1060–1063

Pauling L, Corey RB, Branson HR (1951) The structure of proteins: two hydrogen-bonded helical configurations of the polypeptide chain. Proc Natl Acad Sci U S A 37:205–211

Penrose R (1994) Shadows of the mind. Oxford University Press, Oxford

Perutz MF (1987) I wish I'd made you angry earlier. Scientist 1:9

Richards FM (1958) On the enzymic activity of subtilisin-modified ribonuclease. Proc Natl Acad Sci U S A 44:162–166

Richards FM (1997) Whatever happened to the Fun? An autobiographical investigation. Annu Rev Biophys Biomol Struct 26:1–25

Robertson JM (1938) X-Ray analysis and application of Fourier series methods to molecular structures. Rep Prog Phys 4:332–367

Rossmann MG, Blow DM (1962) The detection of sub-units within the crystallographic asymmetric unit. Acta Crystallogr 15:24–31

Schiltz M, Fourme R, Prangé T (2003) Use of noble gases xenon and krypton as heavy atoms in protein structure determination. Methods Enzymol 374:83–119

Schoeneborn BP, Watson HC, Kendrew JC (1965) Binding of xenon to sperm whale myoglobin. Nature 207:28–30

Smith JMA, Ford GC, Harrison PM, Yariv J, Kalb (Gilboa) AJ (1989) Molecular size and symmetry of bacterioferritin of Escherichia coli: X-ray crystallographic characterization of four crystal forms. J Mol Biol 205:465–467

Sumner JB (1946) The chemical nature of enzymes. Nobel Lecture, 12 December 1946, pp 114–123

Swindells MB, Orengo CA, Jones DT, Hutchinson EG, Thornton JM (1998) Contemporary approaches to protein structure classification. BioEssays 20:884–891

Vijayan M (2001) Obituary: G. N. Ramachandran (1922–2001). Nature 411:544

Yariv J, Kalb AJ, Sperling R, Bauminger ER, Cohen SG, Ofer S (1981) The composition and the structure of bacterioferritin of Escherichia coli. Biochem J 197:171–175

Chapter 2
Tubulin

Colchicine's Claim to Fame

Colchicine, the main alkaloid of *Colchicum autumnale* (of the *Liliaceae* family), was originally extracted from this plant and used in the treatment of gout. To quote Encyclopedia Britannica – "Gout is one of the oldest diseases in medical literature, as colchicine is one of the oldest drugs in therapeutics". But colchicine's claim to fame resides in a quite different achievement to wit, it stops mitotic cell division. The inhibition of cell division by colchicine results in cells that have twice the number of chromosomes than normal cells. Eventually it has been shown that colchicine stops cell division by destroying the structure of the nuclear spindle (Peterson and Mitchison 2002).

The Mitotic Spindle

A protein that is present in the cells of all eucaryotic organisms, mono- and multicellular animals and plants, is the tubulin. This is the protein that forms cytoskeleton, a network of water filled tubules that maintains the shape of the cell. It is also present in cilia and flagella. Temporarily it makes its appearance in the process of cell division as the nuclear spindle that moves the divided chromosomes to the opposite poles of the cell about to divide. The structure of this protein is not known as yet in sufficient detail to explain its properties and the properties of the many structures that it forms in a living cell. It is known for some time now, however, that tubulin is a mixture, in equal proportions, of two similar yet different proteins, the so called α and β tubulins. And indeed the building block of the cytoskeleton, or of cilia or of the nuclear spindle for that matter, is a dimer of some 110,000 daltons, composed of one molecule of α-tubulin and one molecule of β-tubulin. Because the study of tubulin structure is still continuing, even though it has been going on for

© Springer International Publishing Switzerland 2016
J. Yariv, *The Discreet Charm of Protein Binding Sites*,
DOI 10.1007/978-3-319-24996-4_2

some 50 years, it will be instructive to describe some of the thinking and procedures that went into this work.

I should perhaps explain why of all the proteins covered in this book I begin with tubulin which is by far a most difficult object of study and one whose structure has not provided sufficient insights into its functioning in the cell. Perhaps the main reason is that at least in one of its manifestations, the mitotic spindle (Fig. 1.4), it has been in the awareness of biologists for more than a century. Of all the proteins that I deal with in this book, tubulin is the only one that composes structures that can be viewed in a microscope. With all this distinguished background, the story of tubulin does not start till 1960s, when it was realized that quite distinct structures of the cell are all constructed from tubules of about 200 Å diameter and were given the name "microtubules" (Ledbetter and Porter 1963). It is also true that this momentous insight into one of main components of the cell did not receive the attention that it merited, coming at the time that all the rage was the genetic code in all its ramifications.

The 'mitotic-spindle' has a unique position in biological sciences and has already been observed in 1880s. Its discovery coincides with the realization that the chromatin of cell nucleus gives rise to chromosomes before cell division of eucaryotic cells. A landmark in the history of 'mitotic-spindle' is its isolation by Mazia and Dan (1952) (reviewed by Hollenbeck 1985). Discovery of cell skeleton is, of course, more recent depending, as it did, on electron microscopy. With the realization that its main component is the same in cilia and in the mitotic spindle are credited the same Ledbetter and Porter as above. Eventually the protein that makes its presence visible microscopically in various cells and tissues of eucaryotes in the form of microtubules was given the name "tubulin" (Mohri 1968; reviewed by Wells 2005). Beyond the discussion about the origins of tubulin in the eucaryotes (Margulis 1975), the finding of a similar protein in the prokaryotes, the PtsZ-protein, is of special interest (Löwe and Amos 1998; Nogales et al. 1998a). It suggests that, if indeed microtubules are responsible for coordinating the activity of monocellular eucaryotes as suggested by Penrose (1994), then the structural elements composed of PtsZ-protein are similarly responsible for the response of bacteria to external stimuli.

Before the now fashionable proteomics, that is based on the premise that proteins in a living organism do not act as individual molecules but as assemblies, has made its appearance, microtubules were shown to interact with a number of protein entities, the so called "microtubule-associated proteins"(MAPs). All these interesting findings notwithstanding, I shall stick to the limited topic of protein binding-sites to well defined molecules.

A Mine for Tubulin – The Brain

Colchicine is a large hydrophobic compound of the following composition $C_{22}H_{25}NO_6$ (mol. weight 399.4) and its exercises its effect on the nuclear spindle by binding to the tubulin dimer, one molecule of colchicine per dimer. While tubulin dimers can polymerize to form the microtubules of the mitotic spindle, the presence of colchicine in the dimer prevents this polymerization.

The binding of small molecules is a very fundamental property of proteins and one that forms the basis of the many processes that go-on in a living cell. The binding of small molecules to proteins is characterized by specificity and fixed proportions of the interacting molecular species, both properties being the attributes of the so called 'binding-sites'. A binding-site of a protein molecule is a limited region on the protein surface that can bind one molecule of given spatial and chemical characteristics. Before the reality of protein binding-sites was described in structural terms, their existence was concluded from thermodynamic considerations. Of all the different equations describing the reversible small-molecule binding to proteins, the one that has withstood the test of time is the one proposed by Scatchard (Scatchard 1949; Scatchard et al. 1954). In it he formulated four fundamental questions about the binding that has since been quoted often enough. Thus he writes "We want to know of each molecule or ion which can combine with a protein molecule, How many? How tightly? Where? Why?" The advantage in his presentation is that the association constant, K_{ass}, and the so-called "equivalent binding-weight" of the protein can both be readily extrapolated if indeed the binding is described by a single constant. For an appreciation of Scatchard, the man and the scientist, see his obituary in Nature (1974).

The equivalent binding-weight is especially useful if it describes the binding by a pure protein. In such a case the equivalent binding weight bears a simple numerical relation to the molecule weight. Thus, if the equivalent binding-weight equals the weight of the molecule, there is just one binding site to the molecule. If, on the other hand, it equals to half the weight of the molecule this proves that there are two binding sites per molecule and so on. In anticipation of the findings to be related further on suffice is to say that the dimer of tubulin weighs 110,000 daltons, and that it is composed of one monomer of α-tubulin and one monomer of β-tubulin, each weighing 55,000 daltons. Considering that a binding experiment with pure tubulin finds the equivalent binding-weight for colchicine to be appr. 110,000 daltons and that for guanosine-triphosphate (GTP) to be appr. 55,000, it is concluded that both the α-tubulin and the β-tubulin monomers each bind a molecule of GTP and that one molecule of colchicine is bound by the αβ-tubulin dimer. Again, in anticipation of the now available structural information, these conclusions of the binding experiment are confirmed by the solved structure of the complex of tubulin with colchicine (Figs. 2.1 and 2.2) (Ravelli et al. 2004).

Because a tubulin dimer has a binding-site for colchicine, binding of colchicine can be used to monitor the isolation of tubulin from some natural source and as a criterion of the purity of the isolated protein. Even before the truth of the above statements with regard to colchicine was established, colchicine binding was used to monitor the purification of tubulin from pig's brain. Brain was chosen because brain homogenates have high colchicine-binding activity and brain is readily available. Indeed, the isolated tubulin accounted for up to 10 % of the soluble protein of brain homogenates and had properties similar to tubulin isolated from microtubules of cilia and flagella (Weisenberg et al. 1968). With hindsight one can explain why

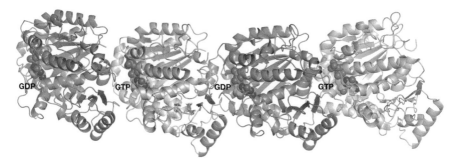

Fig. 2.1 Secondary structure of tubulin dimmers with bound space-filling-models of GDP and GTP (GDP binds to the α–subunit and GTP to β–subunit) (This model was constructed with the data in file 1SA0 in the Protein Data Bank deposited by Ravelli et al. (2004))

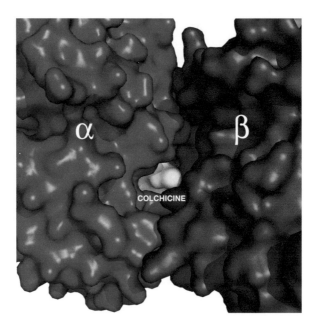

Fig. 2.2 Space-filling-model of the tubulin complex with colchicine (This model was constructed with the data in file 1SA0 in the Protein Data Bank deposited by Ravelli et al. (2004))

this crude method, a rather lengthy filtration used in this work to monitor the binding of colchicine in the various stages of its purification from a brain extract, was successful. Colchicine binds to tubulin with an association constant K_{ass} of 0.42×10^8 l/mol (calculated from Sherline et al. 1975). K_{ass} is also given by the ratio of the on-rate to that of the off-rate of the reacting molecule. When these rates were determined for the binding of colchicine to tubulin it was found that colchicine dissociates from its complex with tubulin at a very slow rate, k_{-1} of 0.009 per hour or, in more friendly terms, that the time it takes for a population of tubulin molecules in complex with colchicine to shed 50 % of colchicine (the famous $t_{1/2}$ term for the half-life) is enormous, being 78 h (Sherline et al. 1975). This lucky coincidence is thus the basis of the success of this crude method in the resulting seminal work on tubulin. The large difference in the rate of association and the rate of dissociation from tubulin suggested that binding of colchicine to tubulin causes a conformational change in the latter (Garland 1978). This claim has now been substantiated with the structure of the colchicine complex with tubulin having been solved (Ravelli et al. 2004). Brain tubulin is now the main object of study of tubulin structure and is available commercially.

Some Insights into Tubulin Structure

The objective of studying a cellular component such as the tubulin is to understand its function. Very often determination of a protein structure to atomic resolution provided an explanation of its function. This was certainly the case in the pioneering studies of the hydrolytic enzymes. This approach was so overwhelmingly successful that solving a protein structure was all that was needed to explain its function. All that was necessary was the availability of a single crystal of the enzyme that diffracted X-rays to a resolution sufficient to discern its atomic structure, usually a resolution of about 2 Å. Situation was not as straightforward when proteins molecules composed of two or more subunits were the object of study, even in the case that the subunits were of one kind. If all that was required to understand the functioning of tubulin in the many cellular structures would be to determine the structure of the dimer, the prospects would not be as daunting as they are at present. As things are, even this limited objective has not been achieved and only recently single crystals of tubulin have been produced that provided information on the conformational changes brought about by the binding of colchicine (Ravelli et al. 2004). The reason could well be the fact that the structure of the tubulin dimer is conditioned to produce extended structures, the microfilaments of which microtubules are composed, rather than the tight 3-D packing that is a prerequisite for producing a crystal. As it is, much of the information about the structure of tubulin antedates this recent X-ray crystallographic study.

To start with, information about the structure of the microtubules came from the electron microscopic study. Thus to quote Ledbetter and Porter (1963) "…. They are of 230–270 Å in diameter in the cortex (200 Å in the spindle) and of undetermined length. The wall of the tubules is about 70 Å thick and the lumen is 100 Å in diameter. ….. There is some evidence that it is made up of smaller filamentous units packed together to form the wall of the cylinder. Presumably, these "smaller units" which appear circular in cross-section represent the major macromolecular element of which the tubules are constructed." Evidence about the "macromolecular element" was provided in a beautiful study by Shelanski and Taylor (1967), carried out on sperm tails of a sea-urchin, that identified it as protein molecule that binds colchicine with a sedimentation constant of 6S. Equally important, even if aesthetically not equally satisfying, was the demonstration that tubulin can re-associate to form microtubules (Weisenberg 1972). Both native microtubules and the re-associated ones became now the object of study by sophisticated methods that utilized optical diffraction of electron micrographs to reconstruct and improve these images by utilizing computer methods and providing a 3-dimensional reconstruction of the original images namely, a 3-dimensional density map (Erickson 1974; Amos and Klug 1974). A summary of these attempts to understand the structure of the microtubules was summarized in a review article by Amos (1975).

Knowledge of the structure of tubuline came from two-dimensional crystalline arrays rather than from a crystal and the method was one of electron – diffraction rather than diffraction of X-rays. This is a much more recent addition to the armoury of protein structure determination than is X-ray diffraction and is immeasurably more tedious. Rather than use the highly automatic methods of data collection now employed at powerful X-ray sources, electron diffraction uses the source of electron microscopes and data analysis is done manually, or at least was until quite recently. Technologically, electron diffraction is in the state that X-ray diffraction of protein crystals was fifty years ago. It was first used successfully in the study of bacteriorhodopsin, a protein pigment in the purple membrane of *Halobacterium halobium*. The study of this protein is memorable for two achievements. It was the first membrane protein whose structure has been solved. It was also the object on which an experiment was performed that convinced the dubious biochemists that the chemiosmotic theory of Mitchell holds for membrane transport, as will be elaborated in Chap. 5. The seminal work on bacteriorhodopsin has been done in San Francisco by Oesterhelt and Stoeckenius (1971) and Blaurock and Stoeckenius (1971) and published in Nature new Biology. Already in 1975 Henderson and Unwin published a paper entitled "Three-dimensional model of purple membrane obtained by electron microscopy" with a memorable illustration (Henderson and Unwin 1975) that was based on a 7 Å resolution map. The demands of this new approach to protein structure solution are evident from the fact that a 3.5 Å resolution map was produced 15 years later and had to use facilities further afield than those available at the Laboratory of Molecular Biology in Cambridge (Henderson et al. 1990).

Electron- diffraction of two-dimensional zinc-induced crystalline sheets of tubulin provided the first 3-dimensional structure of tubulin at 6.5 Å resolution (Nogales et al. 1995). Subsequent work from the same laboratory, The Life Science Division of The Lawrence Berkeley Laboratory, resolved the structure of tubulin to 3.7 Å (Nogales et al. 1998b; reviewed by Pennisi 1998) and to 3.5 Å (Löwe et al. 2001). This work established that each subunit of the dimer of tubulin is constructed of two β-sheets and 12 α-helices (see Fig. 2.1). Most of the known amino-acid sequences could be fitted into the electron density of the two subunits even if assignation which of the subunits is α and which β had to be decided by other considerations. It was also of a quality sufficient to locate the binding-sites of guanosine-triphosphate (GTP) and/or of guanosine-diphosphate (GDP) and the binding site of taxol that was used to stabilize the tubulin sheets. Taxol is one of the many natural products that bind to tubulin. While colchicine prevents microtubule formation, taxol stabilizes them (Peterson and Mitchison 2002). Eventually, image processing of electron micrographs of intact microtubules succeeded in resolving the structure of microtubules to an 8 Å resolution (Li et al. 2002). This allowed to revise structural data about the interactions between the protofilaments that have been observed in the zinc-induced crystal sheets and, to quote the authors,"… provided an improved picture of the interactions between the adjacent protofilaments, which are responsible for microtubule stability ….".

As mentioned above, the achievement of the first single-crystal study of tubulin provided information about the structure of the colchicine binding-site and about the conformational change brought about by this binding in the structure of the αβ–dimer. As can be seen in Fig. 2.2, colchicine is bound between the two subunits. Even if most of it is buried in the β subunit it also interacts with the α subunit. Furthermore, on binding of colchicine the shape of the αβ -dimer changes. Whereas a dimer surface in the protofilaments is straight, in its complex with colchicine it is curved. In this shape the dimer cannot realize the contacts necessary to form the microfilaments and de-polymerization sets in.

References

Amos LA (1975) Structure and symmetry of microtubules. In: Borgers M, de Brahander M (eds) Microtubules and microtubule inhibitors. North-Holland Publishing Company, Amsterdam, pp 21–34

Amos LA, Klug A (1974) Arrangement of subunits in flagellar microtubules. J Cell Sci 14:523–549

Blaurock AE, Stoeckenius W (1971) Structure of the purple membrane. Nat New Biol 233:152–155

Erickson HF (1974) Microtubule surface lattice and subunit structure and observations on reassembly. J Cell Biol 60:153–167

Garland DL (1978) Kinetics and mechanism of colchicine binding to tubulin: evidence for ligand-induced conformational change. Biochemistry 17:4266–4272

Henderson R, Unwin PNT (1975) Three-dimensional model of purple membrane obtained by electron microscopy. Nature 257:28–32

Henderson R, Baldwin JM, Ceska TA, Zemlin F, Beckmann E, Downing KH (1990) Model for the structure of bacteriorhodopsin based on high-resolution electron cryo-microscopy. J Mol Biol 213:899–929

Hollenbeck PJ (1985) Mitotic spindles in isolation. Nature 316:393–394

Ledbetter MC, Porter KR (1963) A "microtubule" in plant cell fine structure. J Cell Biol 19:239–250

Li H, DeRosier DJ, Nicholson WV, Nogales E, Downing KH (2002) Microtubule structure at 8 Å resolution. Structure 10:1317–1328

Löwe J, Amos LA (1998) Crystal structure of the bacterial cell division protein FtsZ. Nature 391:203–206

Löwe J, Li H, Downing KH, Nogales E (2001) Refined structure of αβ-tubulin at 3.5 Å resolution. J Mol Biol 313:1045–1057

Margulis L (1975) Microtubules and evolution. In: Borgers M, de Brabander M (eds) Microtubules and microtubule inhibitors. North-Holland Publishing Company, Amsterdam, pp 3–18

Mazia D (1961) How cells divide. Sci Am 205:100–121

Mazia D, Dan K (1952) The isolation and biochemical characterization of the mitotic apparatus of dividing cells. Proc Natl Acad Sci U S A 38:826–838

Mohri H (1968) Amino-acid composition of "tubulin" constituting microtubules of sperm flagella. Nature 217:1053–1054

Nature (1974) Obituary G. Scatchard 248:367

Nogales E, Wolf SG, Khan IA, Luduena RF, Downing KH (1995) Structure of tubulin at 6.5 Å and location of the taxol-binding site. Nature 375:424–427

Nogales E, Downing KH, Amos LA, Löwe J (1998a) Tubulin and FtsZ form a distinct family of GTPases. Nat struct Biol 5:451–458

Nogales E, Wolf SG, Downing KH (1998b) Structure of the αβ tubulin dimer by electron crystallography. Nature 391:199–203

Oesterhelt D, Stoeckenius W (1971) Rhodopsin-like protein from the purple membrane of Halobacterium halobium. Nat New Biol 233:149–152

Pennisi E (1998) Structure of Key cytoskeletal protein tubulin revealed. Science 279:176–177

Penrose R (1994) Shadows of the mind. Oxford University Press, Oxford

Peterson JR, Mitchison TJ (2002) Small molecules, big impact: a history of chemical inhibitors and the cytoskeleton. Chem Biol 9:1275–1285

Ravelli RBG, Gigant B, Curmi PA, Jourdain I, Lachkar S, Sobel A, Knossow M (2004) Insight into tubulin regulation from a complex with colchicine and a stathmin-like domain. Nature 428:198–202

Scatchard G (1949) The attractions of proteins for small molecules and ions. Ann N Y Acad Sci 51:660–672

Scatchard G, Hughes WL Jr, Gurd FRN, Wilcox PE (1954) The interaction of proteins with small molecules and ions. In: Gurd FRN (ed) Chemical specificity in biological interactions. Academic, New York, pp 193–219

Shelanski ML, Taylor EW (1967) Isolation of a protein subunit from microtubules. J Cell Biol 34:549–554

Sherline P, Leung JT, Kipnis DM (1975) Binding of colchicine to purified microtubule protein. J Biol Chem 250:5481–5486

Weisenberg RC (1972) Microtubule formation in vitro in solutions containing low calcium concentrations. Science 177:1104–1105

Weisenberg RC, Borisy GG, Taylor EW (1968) The colchicine-binding protein of mammalian brain and its relation to microtubules. Biochemistry 7:4466–4479

Wells WA (2005) The discovery of tubulin. J Cell Biol 169:552

Chapter 3
A Variety of Saccharide Binding-Sites

Some Sugar Chemistry

Of all the components of living matter, carbohydrates are chemically the simplest. Most are composed of three elements namely carbon (C), hydrogen (H) and oxygen (O) and most are neutral molecules that is they are neither acids nor bases. Carbohydrates are predominantly polysaccharides. Their building blocks, the monosaccharides, conform to the general formula of the following form $(CH_2O)n$ where n can be as low as 3 to give a triose or as high as 7 to give a heptaose while most components of polysaccharides are pentoses ($n = 5$) or hexoses ($n = 6$). An aqueous solution of a monosaccharide is a mixture, freely interchangeable, of a chain form and of one or a number of ring forms. While the chain form must contain either a chemical grouping defined as an aldehyde or one defined as a ketone, ring forms of carbohydrates have only hydroxyls as functional chemical groupings. Polysaccharides are composed solely from the ring forms of the monosaccharides. This is also true of disaccharides, trisaccharides and higher carbohydrate oligomers. It also holds for saccharides composed of more than one kind of monosaccharide. In the simple case of a disaccharide, depending on whether the so-called reducing-grouping, namely the aldehyde or ketone grouping, of only one or both of the mono-saccharides is utilized in forming the link between the two monosaccharides, the disaccharide is "reducing" or "nonreducing". In the formation of a disaccharide from two monosaccharides two hydroxyl groupings, one from each monosaccha-ride, are utilized to form an ether linkage with unique properties, a glycosidic link-age, between the two units with a release of one molecule of water. The reverse reaction namely, the splitting of the glycosidic linkage between two monosaccha-rides, is hydrolysis and utilizes one molecule of water to reconstitute the two hydroxyls of the original monosaccharides.

Chemical simplicity of carbohydrates notwithstanding, saccharides of even the same chemical formula can have different structures. Because a hexose has five asymmetric carbons it can exist in as many as 32 alternative configurations, the steric isomers of a hexose. Of these, the fifth carbon determines whether a hexose is of D or L configuration. If the hexose is an aldose, namely the potential aldehyde is on carbon 1, configuration of the hydroxyl on carbon 1 determines whether it is alpha (α) or beta (β). Excluding, the configurations on carbon 1 and 5, the eight remaining isomers of a hexose are known by their trivial names like, glucose, galactose etc.

In its ring forms a hexose, again, can exist either as a furanose, a five-member ring, or as a pyranose, a six-membered ring. Both these rings can assume different conformations, especially when bulky groupings are attached to them. While the different conformations, 24 of them, of the unsubstituted hexose in the furanose form are energetically nearly equivalent, a hexose in the pyranose form will exist in one or a few of the energetically preferable conformations the so called chair and boat forms of the molecule. A direct demonstration of their convertibility was realized when a mechanical force was applied to a single molecule and a conversion of a boat to chair conformation within a polysaccharide was observed (Marszalek et al. 1999).

Proteins That Bind Sugars and Proteins That Split Sugars

For a long while, before the prominence of nucleic acids and proteins in biological processes was established, carbohydrates were considered as the sites and agents of biological specificity. Correspondingly, the amount of research that went into the study of the chemistry and biological significance of carbohydrates was immense. The illustrious names of Emil Fischer and Oswald Avery are connected with research into biological specificity and the importance of carbohydrates. Thus, Fischer is credited with comparing the sterical fit of a substrate to an enzyme that splits it to the fit of "a key in a lock" when he studied the splitting of chemically synthesized and sterically defined glycosides by respective enzymes. Avery used carbohydrate markers on Pneumococci envelopes as genetic markers in his epoch defining researches on the genetic function of DNA.

As was mentioned in the 'Introduction', lysozyme was the first enzyme whose structure has been solved by the team headed by Philips (Blake et al. 1965; Phillips 1966) and solution of its structure has suggested a mechanism by which this hydrolytic enzyme splits its saccharide substrate. In this case the substrate is not a neutral saccharide, it contains amine and acetyl substituents, and its binding site is rather extended. Because lysozyme is a relatively small molecule and enzymes that split much smaller substrates, like the β-galactosidase of E. coli, are huge by comparison, it appeared for awhile a paradox. Now with the structures and mechanism of action of a host of hydrolytic glycosidases having been solved, it is clear that all these enzymes, whether big or small, use the same fundamental mechanism in hydrolyzing their respective substrates (Kirby 1996; White et al. 1996; Vocadlo et al. 2001). They all also bind their substrates before splitting them. Alternatively, saccharide-binding proteins with well established function exist that are not enzymes.

Coincidental with crystallization of urease, Sumner had obtained crystals of other pure proteins from the same source, the Jack bean. One of these, concanavalin A, was shown by him to bind certain saccharides. While the biological function of concanavalin A, if any, is unknown to this day, other proteins from a variety of sources were isolated that share the property of binding saccharides. This property is always traced to the saccharide binding-site of the respective protein.

The Saccharide Binding-Site of Concanavalin A

As mentioned already, there is no known biological function connected with concanavalin A and yet, starting in the late 1960s it attracted considerable effort by many research groups to solve its structure. Thus the laboratory of another Nobel Price winner, Gerrard Edelman, working in the Rockefeller Institute, produced a paper entitled "The covalent and three-dimensional structure of concacanavalin A" (Edelman et al. 1972). Without distracting from the seminal importance of this work, its claim as to the location of the saccharide binding-site was proven to be erroneous. The main reason why solution of the structure of concanavalin A in the 1970s failed to locate its saccharide binding-site was that crystals of concanavalin A dissolve on addition of saccharides. By this time it was known from saccharide-binding experiments that a concanavalin A molecule is a dimer composed of two identical monomers and that it binds 2 molar equivalents of methyl α-D-glucopyranoside or, in other words, that each monomer of the protein possesses a saccharide binding-site.

Of the many early attempts to delineate the position of the saccharide binding-site on the surface of the concanavalin A monomer, the one by Brewer et al. (1973) deserves special mention. In this work, the relatively novel ^{13}C Nuclear Magnetic Resonance technique has been utilized to determine the binding of ^{13}C labeled methyl α-D-glucopyranoside and of methyl β-D-glucopyranoside to concanavalin A. The method is based on the fact that a paramagnetic manganese ion residing in the protein, in a well defined site (Weinzierl and Kalb 1971), broadens the resonances of the sugar based ^{13}C-enriched carbons. From the degree of the broadening of the different carbons the distance of the carbons to the manganese ion was determined. It was found that both glucosides were positioned at a distance of 10 Å from the manganese binding-site. Furthermore the measured distances corresponded to both glucosides being in the C1 chair conformation. The unique finding of this work was the fact that the β-glucoside was bound in a reverse orientation to that of the α-glucoside. Reverse binding of ligands is not restricted to saccharide binding-sites, however. Thus, the reverse binding of bromophenols to the tyroxine binding-site of human transthyretin is described and the structural reasons for this reversal in the binding mode are given (Gosh et al. 2000). For awhile it was assumed that the inducer and the anti-inducer were bound to lac-repressor in a reverse orientation, however this is not the case as will be elaborated further on.

Eventually the saccharide binding-site of concanavalin A was solved when crystals of its complexes with methyl α-D-glucopyranoside and methyl α-D-mannopyranoside were produced. Solution of the structure of a cubic crystal of the complex with the glucoside was long in coming (Yariv et al. 1987; Harrop et al. 1996) and the crystal of the complex with mannoside was solved first (Derewenda et al. 1989; Naismith et al. 1994). Electron density of the mannoside bound in the saccharide binding-site of concanavalin A is in Fig. 3.1 (left). Figure 3.1 (right) shows a surface view of the mannoside bound to concanavalin A.

An outstanding property of concanavalin A binding-site and one that was not anticipated is the number of the van der Waals interactions that it makes with the saccharide and the fact that many of these are with the aromatic tyrosine. This feature is common to most saccharide-binding proteins.

The Amazing Cellulose-Binding Protein

Cellulose, the most abundant material produced biologically, which is nothing more than an extended chain of D-glucose units in which the glucohexopyranoses are linked in a β-1-4 linkage, is split by a multitude of the cellulose splitting enzymes. These are of two kinds: the endoglucanases that split internal β-glucosidic linkages and exoglucanases that split such a linkage from the reducing end of the chain to give the disaccharide, glucosyl-β-1-4-glucose, the cellobiose. Cellulose can be highly ordered in the form of crystalline-cellulose and less ordered in the form of amorphous-cellulose or semi-ordered paracrystalline-cellulose. Enzymes that hydrolyse crystalline-cellulose contain distinct non-catalytic cellulose-binding domains, the so-called carbohydrate-binding-domains (CBD), that disrupt the

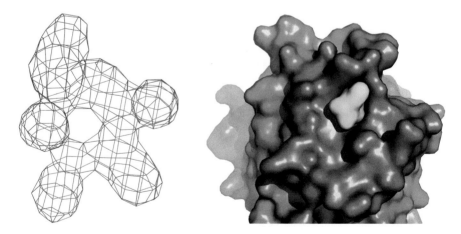

Fig. 3.1 The saccharide binding-site of concanavalin A. One picture showing electron-density of methyl α-D-mannoside stripped of the electron-density of the protein to which it is bound; the other – a surface view of the mannoside bound to concanavalin A

structure of the insoluble crystalline cellulose fibers and thus facilitate its hydrolysis by the enzymes' active sites (Din et al. 1991). Carbohydrate-binding-domains are also present in enzymes of different specificity that hydrolyze other carbohydrate polysaccharides such as starch, xylan and chitin. But even exoglucanases are present in such a variety of forms that they are divided into 28 families (Notenboom et al. 2001).

I have thought it worthwhile to dwell on the structure of CBD of a cellulose splitting enzyme, the Cellobiohydrolase I of *Trichoderma reesei*, for two reasons (Divne et al. 1998). One is the sheer beauty of its binding-site that forms an extensive cellulose-binding tunnel that to quote the authors "was initially estimated to contain seven glucose-binding sites" (Fig. 3.2) The other is that the chemistry of its binding-site resembles that of proteins that bind small saccharides, thus stressing the universality of the chemical principles that are utilized in the binding of saccharides by proteins.

Isolation of a Saccharide-Binding Protein That Controls Gene Expression

It is a matter of opinion but I think that the most important saccharide-binding proteins are the ones that emerged from the study of adaptive enzyme formation in E. coli that has been carried out in the Pasteur Institute in Paris by Jacques Monod and his collaborators. The most revolutionary of these is the repressor whose story is one of the subjects of this chapter. Hardly less significant are the permeases that are

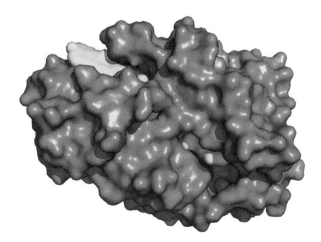

Fig. 3.2 Surface view of cellohexaose bound to cellobiohydrolase I (This figure was constructed with data in files 7CEL and R7CELSF deposited by Divne et al. 1998)

discussed in Chap. 5 and whose existence was brought to the attention of sceptical biochemists in the early fifties by the same group at the Pasteur Institute.

The isolation of lac-repressor by Walter Gilbert and Benno Müller-Hill is a story that has no counterpart in biochemical literature for the audacity and determination with which it was carried out. True, it was preceded by a brilliant analysis of the phenomenon of adaptive enzyme formation in *Escherichia coli* by Francois Jacob and Jacques Monod and a host of outstanding scientists around them who enriched the language with such fundamental new terms as the operon, operator and repressor that they invented to explain the properties of these hypothetical entities and that have become part and parcel of modern genetics terminology. Jacob and Monod did not shy away from devising a new term for the kinetics of the interaction of the hypothetical repressor with the inducer which they named allostery. The concept of allostery became very fruitful in the study of multisubunit proteins, especially enzymes, but the original authors' idea about the chemical nature of the repressor was wrong. They thought that it might be a nucleoprotein but on its isolation it was shown to be a pure protein.

Isolation of lac-repressor started from a premise that there are only a few repressor molecules in an E. coli cell, a premise that is correct. Because the binding constant of the repressor for the inducer IPTG (isopropyl-β-D-thiogalactoside) was calculated to be only 1.7×10^6 l/mol, a tight-binding mutant expected to give a binding constant of 1.4×10^7 l/mol was used as was a radioactive inducer of high specific activity. Having thus made the correct guesses and having correctly interpreted available evidence Walter Gilbert and Benno Müller-Hill (1966) went on to isolate a practically pure repressor by normal run-of-the-mill procedures without any histrionics. In doing so they also proved the premise that they started from namely, that lac-repressor is a pure protein.

lac-Repressor inhibits the expression of lac-operon by binding to lac-operator, a gene that preceeds the sequence of the genes, the so called structural genes, that specify the proteins of this operon. Of these the most famous is the β-galactosidase that splits the disaccharide lactose, or milk-sugar, into D-glucose and D-galactose. lac-Repressor is also a product of a structural gene but one that is not part of lac-operon but precedes it in the DNA sequence of E. coli chromosome. Some mutations of this gene give repressor molecules that bind inducer more strongly than the wild-type repressor while some other mutations give repressor that has lost the capacity to bind inducer.

The inducer itself is not a single compound but a member of a family of compounds. When prepared synthetically, the anomeric oxygen that connects the galactopyranosyl entity to the aglycon can be replaced by an atom of sulphur. Historically, these compounds were prepared to prevent their splitting by the ever present β-galactosidase. Once it was found that some of these thiogalactosides can induce the production of β-galactosidase, they were used extensively.

Isolation of lac-repressor was preceded by a rather crude study of its binding properties which showed that among the many glycosides tested a few, rather than release the hypothetical binding of repressor to operator, caused a tighter interaction between repressor and operator (Müller-Hill et al. 1964). Eventually one of these compounds was used to strengthen the interaction of repressor with DNA in attempts to crystallize it (Pace et al. 1990; Bell and Lewis 2001). An interesting observation in this early success of producing co-crystals of lac-repressor with lac-operator was that they crack on being exposed to inducer. A similar behavior was known to occur with crystals of reduced haemoglobin on exposure to air, the cause being that the structures of haemoglobin and oxyhaemoglobin are different (Perutz et al. 1964) An observation that came too late to be utilized in structure solution of lac-repressor was the amazing finding that changing a single lysine, Lys84, in the sequence of lac-repressor for methionine produced a heat-stable repressor. While wild-type repressor denatures when exposed to temperatures above 53 °C, the Met84 mutant could resist exposure to temperatures of 93 °C (Pereg-Gerk et al. 2000) Thus, lac-repressor is afforded a new lease-on-life as an object for the study of protein heat stability. As could have been expected, heat-stable repressor produces better refracting crystals than the wild-type repressor (Bell et al. 2001). Using a mutant of lac-repressor that produces dimers, instead of the tetramers produced by the wild-type, and replacing Lys84 with leucine gave crystals that diffracted to 1.7 Å. With lysine in position, isomorphous crystals of the dimer-producing mutant diffracted to only 3.0 Å.

Binding of lac-Repressor to DNA – A Molecular Acrobatics

Today when sequences of DNA stretches of interest are done routinely in medical laboratories, it is difficult to appreciate the seminal role that the invention of sequencing had in the advance of molecular biology. The father of it all was Frederick Sanger who produced the first sequence of a protein, the insulin, by a rather primitive but ingenious method and was awarded a Nobel Prize for it. Rather than rest on his laurels, he then turned to devising a method for sequencing DNA. Again he was successful and again he was awarded a Nobel Price for it but he shared it with Walter Gilbert who devised a different method of DNA sequencing. The mediator for Gilbert's Nobel was the lac-repressor that he isolated with Müller-Hill and whose interaction with DNA, the lac-operator, he studied (Gilbert and Maxam 1973).

lac-Operator and other operators discovered since have proven to be more sophisticated than the concept of a gene that existed then. There is a multiplicity of separate binding sites for lac-repressor on the DNA rather than a single site. The credit for it must go primarily to Müller-Hill who has shown by judicious use of mutants that releasing of the binding of lac-repressor to operator O_1 in the presence of inducer increases β-galactosidase production only 20-fold whereas in the wild strain, with all the sites available, it increases 1000-fold. Releasing the binding to either operator O_2 or operator O_3 increases β-galactosidase production even less than to operator O_1. On top of this, he also showed that a mutant producing only dimers increased β-galactosidase production in presence of inducer only slightly whereas wild-type producing tetramers increased it fully (Oehler et al. 1990). Because the location of these three operators on the DNA has been mapped, it has

34   3 A Variety of Saccharide Binding-Sites

been suggested that lac-repressor molecule can bind to two operators simultaneously (Kania and Müller-Hill 1977) and thus form a loop of the DNA attached to a molecule of lac-repressor. This part of lac-repressor fact and guess-work has now been substantiated as can been seen in Fig. 3.3 reproduced from Lewis et al. (1996).

Its Structure

Structure solution of lac-repressor was slow in coming, the reason being that no single crystals of lac-repressor of adequate quality could be produced and it was not for lack of trying. When eventually such crystals were produced the results were published under the impressive title Crystal Structure of the Lactose Operon Repressor and Its Complexes with DNA and Inducer (Lewis et al. 1996) and reviewed by Matthews in the same issue of Science. By then it was shown that intact lac-repressor molecule is cleaved by the action of trypsin into four molecular

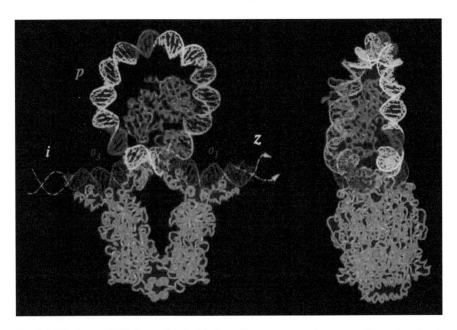

Fig. 3.3 The loop of DNA formed by the binding of lac-repressor to lac-operon (Reproduced with permission from Fig. 11 of Lewis et al. 1996). A model of the 93-bp repression loop that corresponds to the lac operon-82 to +11. The ends of the loop contain operators 01 and 0.3, (*red*) to which the lac repressor tetramer is shown bound (*violet*). Inserted in the loop is the CAP protein and 30-bp DNA complex (*blue*) taken from the PDB coordinates. The grey DNA was created by applying a smooth curvature to B-DNA

fragments of low molecular weight that correspond to the amino-terminal of the molecule (N-terminal sub-domain) and a high-molecular tetrameric core fragment that represents the rest of the molecule (the four C-terminal sub-domains). While the low-molecular fragment binds to DNA, the core contains the inducer binding-sites.

The main interest in these studies was and continues to be the interaction of lac-repressor with DNA and the allosteric conformational change brought about by its interaction with the inducer that releases the binding of the repressor to DNA. However the quality of data of this publication and the one preceeding it from Yale that gave the structure of the lac-repressor core (Friedman et al. 1995) were not satisfactory or sufficient to provide an explanation of this phenomenon. Eventually a crystal of lac-repressor with bound ITPG was solved to a resolution of 2 Å that provides the structure of the saccharide binding-site of lac-repressor at atomic reso-lution (Fig. 3.4) as well as an explanation of the structural changes of the molecule on binding of an inducer that release it from binding to DNA, the operator (Daber et al. 2007). It also corrects a previous conclusion that IPTG and ONPF (orthonitrophenyl-β-D-fucoside) bind to lac-repressor in a reverse orientation. Surprisingly, the structure of repressor with bound ONPF is very similar to the structure of the IPTG-repressor complex. There is however a difference between the structure of the free repressor and the IPTG-repressor complex. Binding of IPTG creates an extensive water-mediated network that crosslinks the N and C-terminal sub-domains of the repressor subunit. As stated above the saccharide-binding site of repressor resides in the C-terminal sub-domain and indeed with the exception of O6 all the other oxygen atoms of the galactosyl residue form hydrogen bonds with amino-acid side-chains of the C-terminal sub-domain. Furthermore, this saccharide

Fig. 3.4 Surface view of IPTG bound to lac-repressor (This figure was constructed with data in file 2P9H in the Protein Data Bank deposited by Daber et al. 2007)

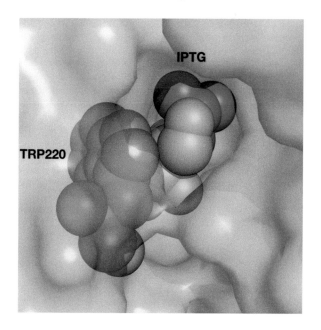

binding-site possesses a hydrophobic pocket consisting of isoleucine (I), leucine (L) and two residues of phenylalanine (F) in the C-terminal sub-domain. In the IPTG-repressor comlex, this is the pocket in which the isopropyl residue nestles. By analogy this is also where the methylene residue of allolactose (galactosyl-β–1–6–glucose) must reside, confirming the importance of van-der-Waals interactions in saccharide binding to protein sites.

Anticipating the tenet of the allostery model discussed in the next chapter that assumes at least two interchangeable conformations of a multimeric protein, one would have expected the structures of the IPTG and ONPF complexes of the repressor to be different considering that IPTG releases lac-repressor from its interaction with the lac-operator while ONPF stabilizes it. The reality in this case is that indeed repressor conformation changes in the process of releasing it from its binding to the operator by binding IPTG, but this happens in the presence of the operator. To quote Daber et al. (2007), "In the presence of the operator the repressor adopts a conformation that is significantly different from the apo repressor or the repressor bound to effector molecules". And further on, "This structural rearrangement alters both the intramolecular interactions of the monomer and the intermolecular interactions between the two N-terminal subdomains".

References

Bell CE, Lewis M (2001) Crystallographic analysis of Lac repressor bound to natural operator O1. J Mol Biol 312:921–926

Bell CE, Barry J, Matthews KS, Lewis M (2001) Structure of a variant of lac repressor with increased thermostability and decreased affinity for operator. J Mol Biol 313:99–109

Blake CCF, Koenig DF, Mair GA, North AC, Phillips DC, Sarma VR (1965) Structure of hen egg-white lysozyme. A three-dimensional Fourier synthesis at 2 Å resolution. Nature 206:757–761

Brewer CF, Sternlicht H, Marcus DM, Grollman AP (1973) Interaction of saccharides with concanavalin A. Mechanism of binding of α- and β-methyl D-glucopyranoside to concanavalin A as determined by ^{13}C nuclear magnetic resonance. Biochemistry 12:4448–4457

Daber R, Stayrook S, Rosenberg A, Lewis M (2007) Structural analysis of lac repressor bound to allosteric effectors. J Mol Biol 370:609–619

Derewenda Z, Yariv J, Helliwell JR, Kalb(Gilboa) AJ, Dodson EJ, Papiz MZ, Wan T, Campbell J (1989) The structure of the saccharide binding site of concanavalin A. EMBO J 8:2189–2193

Din N, Gilkes NR, Tekant B, Miller RCJ, Warren RAJ, Kilburn DG (1991) Non-hydrolytic disruption of cellulose fibers by the binding domain of a bacterial cellulase. Biotechnology 9:1096–1099

Divne C, Ståhlberg J, Teeri TT, Jones A (1998) High-resolution crystal structures reveal how a cellulose chain is bound in the 50 Å long tunnel of cellobiohydrolase I from trichoderma reesei. J Mol Biol 275:309–325

Edelman GM, Cunningham BA, Reeke GN Jr, Becker JW, Waxdal MJ, Wang JL (1972) The covalent and three-dimensional structure of concanavalin A. Proc Natl Acad Sci U S A 69:2580–2584

Friedman AM, Fischmann TO, Steitz TA (1995) Crystal structure of lac repressor core tetramer and its implications for DNA looping. Science 268:1721–1727

Gilbert W, Maxam A (1973) The nucleotide sequence of the lac operator. Proc Natl Acad Sci U S A 70:3581–3584

Gilbert W, Müller-Hill B (1966) Isolation of the lac repressor. Proc Natl Acad Sci U S A 56:1891–1898

Gosh M, Meerts IATM, Cook A, Bergman A, Brouwer A, Johnson LN (2000) Structure of human transthyrethin complexed with bromophenols: a new mode of binding. Acta Crystallogr D Biol Crystallogr D56:2189–2193

Harrop SJ, Helliwell JR, Wan TCM, Kalb(Gilboa) AJ, Tong L, Yariv J (1996) Structure solution of a cubic crystal of concanavalin A complexed with methyl α-D-glucopyranoside. Acta Crystallogr D Biol Crystallogr D52:143–155

Kania J, Müller-Hill B (1977) Construction, isolation and implications of repressor-galactosidase – β-galactosidase hybrid molecules. Eur J Biochem 79:381–386

Kirby AJ (1996) Illuminating an ancient retainer. Nat Struct Biol 3:107–108

Lewis M, Chang G, Horton NC, Kercher MA, Pace HC, Schumacher MA, Brennan RG, Lu P (1996) Crystal structure of the lactose operon repressor and its complexes with DNA and inducer. Science 271:1247–1254

Marszalek PE, Pang Y-P, Li H, ElYazal J, Oberhauser AF, Fernandez JM (1999) Atomic levers control pyranose ring conformations. Proc Natl Acad Sci U S A 96:7894–7898

Matthews KS (1996) The whole lactose repressor. Science 271:1245–1246

Müller-Hill B, Rickenberg HV, Wallenfels K (1964) Specificity of the induction of the enzymes of the lac operon in Escherichia coli. J Mol Biol 10:303–318

Naismith JH, Emmerich C, Habash J, Harrop SJ, Helliwell JR, Hunter WN, Raftery J, Kalb(Gilboa) AJ, Yariv J (1994) Refined structure of concanavalin A complexed with methyl α-D-mannopyranoside at 2 Å resolution and comparison with saccharide-free structure. Acta Crystallogr D Biol Crystallogr D50:847–858

Notenboom V, Boraston AB, Chiu P, Freelove ACJ, Kilburn DG, Rose DR (2001) Recognition of cello-oligosaccharides by a family 17 carbohydrate-binding module: an X-ray crystallographic, thermodynamic and mutagenic study. J Mol Biol 314:797–806

Oehler S, Eismann ER, Krämer H, Müller-Hill B (1990) The three operators of the lac operon cooperate in repression. EMBO J 9:973–979

Pace HC, Lu P, Lewis M (1990) Lac repressor: crystallization of intact tetramer and its complexes with inducer and operator DNA. Proc Natl Acad Sci U S A 87:1870–1873

Pereg-Gerk L, Leven O, Müller-Hill B (2000) Strengthening the dimerization interface of lac repressor increases its thermostability by 40 °C. J Mol Biol 299:805–812

Perutz MF, Bolton W, Diamond R, Muirhead H, Watson HC (1964) Structure of haemoglobin, an X-ray examination of reduced horse haemoglobin. Nature 203:687–690

Phillips DC (1966) Sci Am 215:78–90

Vocadlo DJ, Davies GJ, Laine R, Withers SG (2001) Catalysis by hen egg-white lysozyme proceeds via a covalent intermediate. Nature 412:835–838

Weinzierl I, Kalb AJ (1971) The transition metal-binding site in concanavalin A at 2.8 Å resolution. FEBS Lett 18:268–270

White A, Tull D, Johns K, Withers SG, Rose DR (1996) Crystallographic observation of a covalent catalytic intermediate in a β-glycosidase. Nat Struct Biol 3:149–154

Yariv J, Kalb(Gilboa) AJ, Papiz MZ, Helliwell JR, Andrews SJ, Habash J (1987) Properties of a new crystal form of the complex of concanavalin A with methyl α-D-glucopyranoside. J Mol Biol 195:759–760

Chapter 4
The Secret of Protein Sophistication

Protein Beauty – Protein Complexes with Chromophores

A special kind of a protein binding-site is that found in the so-called holoproteins that are stable complexes of proteins with nonprotein small molecules, the so-called chromophores. As the name indicates these small molecules are responsible for the color of the holoproteins found in nature. To name but a few: the red haemoglobin; the green chlorophyll; the purple color of the purple bacteriorhodopsin that has played such a central role in the study of membrane proteins and membrane transport. When the chromophore is removed from such proteins the resulting apo-proteins are as unattractive to look at as the proteins that we have been dealing with until now to say, they are colourless. The processes developed for the removal of the chromophores from the holoproteins eventually reached a degree of sophistication that allowed the researchers to reconstitute the original protein by mixing the chromophore with the isolated apo-protein (Teale 1959). The resultant reconstituted proteins have been shown to possess the same chemical and physiological properties as the holoproteins isolated from biological material. Perfidiously, the scientists could now cheat the innocent apo-proteins to accept chromophores with an altered chemistry in order to test certain assumptions about their properties (Kitagawa et al. 1982; Albani and Alpert 1987). Our old acquaintance the myoglobin, having been the first protein whose structure has been solved, was the main object of such studies.

Protein Beauty – Symmetry of Multisubunit Proteins

Aesthetically, the one saving grace that the colourless proteins have is the symmetry exhibited by protein molecules composed of a number of subunits. Thus, the lac-repressor molecule described in the previous chapter is tetramer of a point 2,2

© Springer International Publishing Switzerland 2016
J. Yariv, *The Discreet Charm of Protein Binding Sites*,
DOI 10.1007/978-3-319-24996-4_4

symmetry. The Aquaporin and the Voltage Dependent Potassium Channel that will be described in the next chapter are tetramers where the four identical subunits of these molecules are fourfold symmetrical. Haemoglobin is also a tetramer of a point 2 symmetry, except that the dimer is composed of two similar but different subunits, the α-subunit and the β-subunit.

A Most Beautiful Theory – Allostery

The term allostery preceded the model devised to explain phenomena observed by a host of biochemists and protein chemists who studied interactions of proteins with small molecules or, alternatively, the metabolism of nutrients. It was coined by Jacob and Monod to explain the mechanism by which an inducer releases the repressor of lac-operon from its interaction with lac-operator as described in Chap. 3. This then was expanded to a most beautiful model (known as the MWC model) that has revolutionized the way we think about a multitude of phenomena (Monod et al. 1965). The beauty of this idealized model rests on a number of assumptions and its consequences are depicted mathematically. It is a result of collaboration between Monod and his student Changeux, biochemists specializing in metabolic phenomena, with Wyman, a physical chemist with experience in the intricacies of the mechanism of the binding of oxygen to haemoglobin. John T. Edsall, a protein chemist of renown, writing in 1990 about Wyman entitled his article "Jeffries Wyman: Scientist, Philosopher and Adventurer", and Maurizio Brunori writing about Wyman in 1999 entitled his article in the Reflections section of TIBS "Hemoglobin is an honorary enzyme" adapting the witticism expressed by Wyman and Allen (1951), to wit ... "if we are prepared to accept hemoglobin as an enzyme, its behaviour might give us hint as to the kind of process to be looked for in enzymes more generally". The culmination of Wyman's contribution to allosteric theory is his book coauthored with Gill in 1990 and entitled "Binding and Linkage: Functional Chemistry of biological Macromolecules".

MWC model is valid for multi-subunit proteins, oligomers, where each subunit, a protomer, possesses one kind of site. The theory utilized symmetry considerations of multi-subunit proteins to propose that the observed phenomena, such as the cooperative binding curves, were due to the fact that the molecule can exist in two conformations, relaxed (R) and tense (T), that are in equilibrium and that the affinity of the binding-sites in the two conformations is very different. Binding of the small molecule, the effector, to a single site of the T form affects this equilibrium by stabilizing the molecule in the alternative R conformation. In allosteric enzymes two kinds of stereospecific sites are assumed, the site that binds substrate (S) and the site that binds the effector (F), and both sites have different affinities towards the T and R states. Thus the presence of F will modify the apparent affinity for S and vice-versa. There is a different kind of an allosteric enzyme where there is no difference in the affinity of S for the two states that differ in their catalytic activity. In such a case depending on whether F has maximum affinity for the active or inactive state of the enzyme, binding of F will activate or inhibit the enzyme. The corollary of this model is that co-operative homotropic interactions, as evidenced by the binding curves of the F or S, will only be observed when both ligands have affinity that is different for the R and T states of the protein. The model then goes on to describe the modes utilized by protomers when they associate to form stable oligomeric proteins. This is illustrated in a figure reproduced from the above article (Fig. 4.1) that distinguishes between a heterologous association where the bonding between the protomers is made up of two different binding sets and an isologous association where the bonding involves identical binding sets.

The physical basis of the model is the fact that protein structures are not static. Individual atoms and stretches of the protein chain undergo movements that can result in different conformations of the protein structure. Thus the static structure that emerges from X-ray diffraction studies is an average structure. Occasionally such a study yields a number of conformers of the protein molecule (Smith et al. 1986).

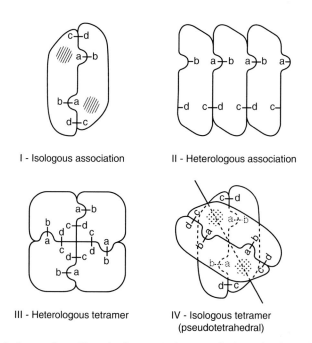

I - Isologous association

II - Heterologous association

III - Heterologous tetramer

IV - Isologous tetramer (pseudotetrahedral)

Upper left: an isologous dimer. The axis of symmetry is perpendicular to the plane of the Figure.
Upper right: "infinite" heterologous association.
Lower left: "finite" heterologous association, leading to a tetramer with an axis of symmetry perpendicular to the plane of the Figure.
Lower right: a tetramer constructed by using isologous associations only. Note that two different domains of bonding are involved.

Fig. 4.1 Isologous and heterologous associations between protomers (This Figure and the text is reproduced with permission from Figure 8 of Monod et al. 1965, p. 107)

Not surprisingly, the theory explained very well the sigmoidal binding of oxygen to haemoglobin in terms of a homotropic allosteric interaction of the model. In Fig. 4.2 reproduced from Wyman and Gill (1990) three principal ways of binding a small molecule by a macromolecule are shown for the binding of oxygen by myoglobin and haemoglobin: (1) by plotting saturation of the molecules by oxygen versus partial oxygen pressure (Fig. 4.2a), (2) by again plotting the saturation but this time versus logarithm of the partial pressure (Fig. 4.2b) and (3) the Hill plot (Fig. 4.2c). To quote the authors "A more rational way of plotting the oxygen-binding data is a function not of the partial pressure itself but of its logarithm....... We shall reserve the term *binding curve* for such plots. In these plots we see that the oxygen binding curve is everywhere steeper for haemoglobin than it is for myoglobin. The relative steepness of the binding curve is brought out in the Hill plot .... ".

The observed cooperative binding of oxygen by haemoglobin was explained by a transition between the two different conformations of the molecule that did not necessitate the assumption of site-site interaction assumed in previous models (Edsall 1980). The assumptions of the model were well confirmed by what was known of the

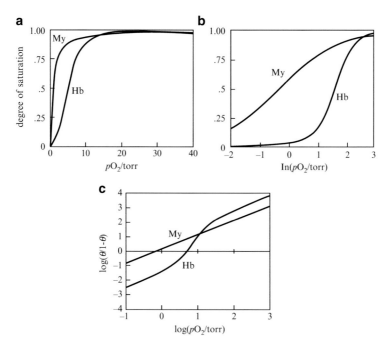

Fig. 4.2 Comparison of oxygen-binding curves of myoglobin and haemoglobin (**a**) A plot of saturation of the molecules by oxygen versus partial oxygen pressure; (**b**) A plot of saturation versus logarithm of the partial pressure; (**c**) The Hill plot (Figure and text reproduced with permission from Figure 1.1 of Wyman and Gill 1990, p. 3)

haemoglobin molecule. Thus it was known from the work of Perutz that the haems in this tetrameric molecule were quite distant from each other (Perutz et al. 1964 and Fig. 4.3). The fact that haemoglobin (T conformation) and oxyhaemoglobin (R conformation) were different was proven much earlier and dramatically when it was observed that crystals of haemoglobin crack when exposed to oxygen (Haurowitz 1938). Eventually this was proven in atomic detail (Perutz et al. 1964).

The correspondence of the MWC model to what was known about the properties of haemoglobin molecule could not be extended, at this stage, to proteins involved in metabolism that showed cooperative binding of the effector e.g., to lac-repressor. And there was, at this time, a host of such observed cases most of which belonged to the so-called end-product inhibition (Umbarger 1961). A unique position in the allostery armoury belongs to an enzyme that splits glycogen, glycogen phosphorylase-b. It was the first allosteric enzyme to be isolated by the husband-and-wife team (Cori and Cori 1936) and shown to be activated by adenine-monophosphate (AMP). Eventually its structure has been solved in the R configuration and compared with the T configuration solved previously (Barford and Johnson 1989).

End-product inhibition of biosynthetic enzyme sequences was a quite well known. phenomenon. Edwin Umbarger, one of the first biochemist to describe it, lists twelve such cases in an article he contributed to cold Spring Harbour Symposia of Quantitative Biology, entitled "Feedback Control of Endproduct Inhibition" (Umbarger 1961). It also figures prominently in an article that predated the MWC model and where the concepts "Allosteric Proteins as Metabolic Regulators" and "Allosteric Effects as Conformational Alterations" were first introduced (Monod et al. 1963).

A good example of an enzyme that conforms to the MWC model is that of aspartate transcarbamylase, in the sense that it is a multimeric molecule composed of two kinds of subunits namely, a large one, the catalytic subunit that binds the substrate, and a small one that binds the effectors. This is an enzyme that catalyses the

Fig. 4.3 A schematic structure of the haemoglobin molecule (Figure is reproduced with permission from fig. 1 of Eaton et al. 1999 who adapted it from Dickerson and Geis 1983)

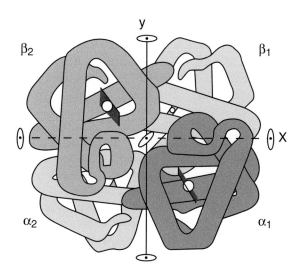

conversion of carbamoylphosphate and aspartate into N-carbamoylaspartate that is the first step in pyrimidine biosynthesis. The end product of the pyrimidine pathway, cytidine triphosphate or CTP, binds to the small subunit and causes a decrease in the catalytic velocity of this enzyme. To quote from an article of Gouaux and Lipscomb (1988), "As the paradigmatic regulatory enzyme, ACTase has been studied for more than 30 years not only in an effort to understand its allosteric properties but also because it catalyses a reaction that is requisite for cell division". X-Ray crystallographic study of the enzyme commences with the work of Wiley and Lipscomb (1968) that established the enzyme morphology and the symmetry of its 12 polypeptide chains that are of two kinds. To quote from this seminal paper "Determination ... of a trigonal crystal form of the CTP-ATC-ase complex shows that ACT-ase lies on a threefold axis in the crystal. Similar investigation on a tetragonal crystal form of ATCase shows that the molecule is situated on a crystallographic twofold axis. The first result means that the number of identical subunits must be a multiple of three, while the second result means that the number must be a multiple of two. The number of equivalent subunits must therefore be a multiple of six. …… Furthermore, if we assume that the symmetry of the ACTase molecule is the same in the trigonal CTP-ACTase crystal as it is in the tetragonal ACTase crystal, the point group symmetry of the molecule must be D_3-32". This thus corrected previous estimates of stoichiometry of the two kinds of subunit weighing 100,000 daltons and 34,000 daltons, isolated by Gerhart and Schachman (1965) when they dissociated the enzyme by treatment with a mercurial reagent. They found that while one of the subunits, the larger one, was catalytically active the other, the smaller, bound CTP. The crystallographic analysis established that the enzyme is composed of two catalytic subunits and three regulatory subunits. The catalytic subunit is composed of three monomers, each weighing 33,000 daltons, and the regulatory subunit is composed of two monomers, each weighing 17,000 daltons (Fig. 4.4).

While the catalytic activity of the isolated catalytic subunit varies hyperbolically with substrate concentration, catalysis by the enzyme varies sigmoidally with substrate concentration. According to the MWC model the explanation of the sigmoidal kinetics derives from the fact that the enzyme exists in two conformations T and R that differ in their affinity for the substrate and in their catalytic rate. Solution of the structures of this enzyme without a ligand in the substrate binding site, the assumed T conformation, and with molecules that stand for the substrate bound in the catalytic site, representing the R conformation, has shown that indeed there is a difference in the structures between T and R even when only one substrate binding-site out of the molecule's six is occupied. This was indeed a superb confirmation of the MWC model prediction of the mechanism of a homotropic allosteric interaction. Not so with the model's prediction of the mechanism of heterotropic allosteric interaction. Crystallography of various crystals of this enzyme could not prove unequivocally that the heterotropic effect operates in the binding of CTP to the effector binding site in the small subunits of the enzyme.

I omit here to mention the tremendous amount of work and the different methodologies that have been used by various groups in studying aspartate transcarbamylase. The structure of this enzyme was solved by William N. Lipscomb and his collaborators.

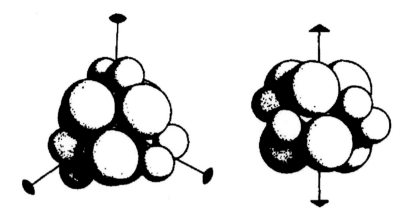

Fig. 4.4 The symmetry of aspartate transcarbamylase molecule (Reproduced with permission from Figure 1 of Kantrowitz et al. 1980 TIBS). Views along the threefold (*left*) and the twofold (*right*) axes of the aspartate transcarbamylase molecule. Catalytic and regulatory monomers are indicated by a large and small ball respectively. Each regulatory dimer connects catalytic chains of different trimers in such a way that the c:r –r:c unit is slanted relative to the threefold axis. It seems plausible that this slant becomes smaller when the c_3 trimers move apart as the substrates bind

This work that stretched over a number of years established that the activity of this enzyme conforms to the MWC model and was summarized in a review entitled "Escherichia coli aspartate transcarbamoylase: the molecular basis for a concerted allosteric transition" (Kantrowitz and Lipscomb 1990).

References

Albani J, Alpert B (1987) Fluctuation domains in myoglobin. Eur J Biochem 162:175–178

Barford D, Johnson LN (1989) The allosteric transition of glycogen phosphorylase. Nature 340:609–616

Brunori M (1999) Hemoglobin is an honorary enzyme. Trends Biochem Sci 24:158–161

Cori CF, Cori GT (1936) Mechanism of formation of hexosemonophosphate in muscle and isolation of a new phosphate ester. Proc Soc Exp Biol Med 34:702–705

Eaton WA, Henry ER, Hofrichter J, Mozzarelli A (1999) Is cooperative oxygen binding by hemoglobin really understood? Nat Struct Biol 6:351–357

Edsall JT (1980) Hemoglobin and the origins of the concept of allosterism. Fed Proc 39:226–235

Edsall JT (1990) Jeffries Wyman: scientist, philosopher and adventurer. Biophys Chem 37:7–14

Gerhart JC, Schachman HK (1965) Distinct subunits for the regulation and catalytic activity of aspartate transcarbamylase. Biochemistry 4:1054–1062

Gouaux JE, Lipscomb WN (1988) Three-dimensional structure of carbamoyl phosphate and succinate bound to aspartate carbamoyltransferase. Proc Natl Acad Sci U S A 85:4205–4208

Haurowitz F (1938) Das Gleichgewicht zwischen Hamoglobin und Sauerstoff. Z Physiol Chem 254:266–274

Kantrowitz ER, Lipscomb WN (1990) Escherichia coli aspartate transcarbamoylase: the molecular basis for a concerted allosteric transition. Trends Biochem Sci 15:53–59

Kantrowitz ER, Pastra-Landis SC, Lipscomb WN (1980) E. coli aspartate transcarbamylase: part II: structure and allosteric interactions. Trends Biochem Sci 5:150–153

Kitagawa T, Ondrias MR, Rousseau DL, Ikeda-Saito M, Yonetani T (1982) Nature 298:869–871

Monod J, Changeux J-P, Jacob F (1963) Allosteric proteins and cellular control systems. J Mol Biol 6:306–329

Monod J, Wyman J, Changeux J-P (1965) On the nature of allosteric transitions: a plausible model. J Mol Biol 12:88–119

Perutz MF, Bolton W, Diamond R, Muirhead H, Watson HC (1964) Structure of haemoglobin, an X-ray examination of reduced horse haemoglobin. Nature 203:687–690

Smith JL, Hendrickson WA, Honzatko RB, Sheriff S (1986) Structural heterogeneity in protein crystals. Biochemistry 25:5018–5027

Teale FWJ (1959) Cleavage of the haem-protein link by acid methylethylketone. Biochim Biophys Acta 35:543

Umbarger HE (1961) Feedback control by endproduct inhibition. Cold Spring Harb Symp Quant Biol 26:301–312

Wiley DC, Lipscomb WN (1968) Crystallographic determination of symmetry of aspartate trans-carbamylase. Nature 218:1119–1121

Wyman JRJ, Allen DW (1951) The problem of the heme interactions in hemoglobin and the basis of the Bohr effect. J Polym Sci 7:499–518

Wyman J, Gill SJ (1990) Binding and linkage, functional chemistry of biological macromolecules. University Science Books, Mill Valley, CA, USA

Chapter 5
Curiouser Binding-Sites – In Membrane Transport Proteins

Membrane Proteins – Sentries at the Gate

Biological membranes were for a long time considered to consist solely of lipid namely, a 30 Å thick bilayer of phospholipid, hydrophobic in its interior (Davson and Danielli 1943). While a biological membrane was considered to be impermeable to cell constituents, it was assumed to be permeable to water and to a lesser degree to solutes. Indeed, for a long time it was known as 'a semipermeable membrane'. Various facets of membrane behaviour confirmed this picture of its structure. Thus cells enclosed by membrane were shown to contract in hypertonic solutions of sucrose and to swell, or even to burst, in hypotonic solutions of this sugar. Of necessity, this simple picture of biological membranes could not be sustained once study of metabolism began in earnest in the twentieth century.

The discovery of cytochrome by Keilin (1925) marked the beginning of the understanding of cell respiration and the pivotal role of mitochondria in cell metabolism. Beside it being a tribute to Keilin the man, the article entitled "Keilin, Cytochrome, and the Respiratory Chain", contributed by Slater (2003a) to JBC Centennial, is a comprehensive account of the many-pronged attack on what was called "the respiratory chain" or alternatively, "the electron transfer chain". Thus, it summarizes the outstanding discoveries before the second world-war, and the illustrious names associated with it, such as the activation of hydrogen by the various dehydrogenases and the transfer of the electrons thus released to an oxygen-activated oxidase by the cytochromes. He then goes on to describe the later achievements such as the discovery of ubiquinone and of iron-sulphur proteins. The review includes discussion of oxidative phosphorylation, of the structure of the mitochondrion and its inner membrane where the cytochromes, with the exception of the soluble cytochrome c, are located. It concludes by praising Mitchell for his contribution to the understanding of the electron transfer and its linkage to the transfer of protons in the opposite direction by the mitochondrial membrane according to his chemi-osmotic hypothesis and his, that is Slater's, regret that he did not heed the

© Springer International Publishing Switzerland 2016
J. Yariv, *The Discreet Charm of Protein Binding Sites*,
DOI 10.1007/978-3-319-24996-4_5

advice of Mitchell with whom he was in correspondence. It should be pointed out that Slater joined the Molteno Institute in 1946 as a Ph.D. student of Keilin and thus was a contemporary of Mitchell in Cambridge. His review is thus autobiographical in part and a good-read.

Even more directly concerned with transport through the semipermeable membrane was the demonstration by Hodgkin and Huxley (1952) of solute specific pores in the membrane of the giant axon of squid when they studied electrical current convection by nerves. But the Jocker in the pack was certainly the demonstration by a series of papers from the Pasteur Institute in Paris of the adaptive formation of a lactose transporting protein in the bacterium E. coli (Rickenberg et al. 1956).

The paradox of the semipermeable membrane are the aquaporins. While everybody working on the fundamental problems of membrane transport proteins was happy to assume that biological membranes were permeable to water, as indeed they are, some people were not, or not quite. These were the physiologists who were studying water balances in animals. They are aware of some balances that ordinary people aren't for example, that kidneys process some 180 l of filtrate a day and that most is resorbed in the kidney and only 1 l is excreted as urine. And, they came to the conclusion that lipid-bilayer-membranes of kidneys, the glomeruli, could not handle such a volume of water-traffic and that there must be ports in these membranes that facilitate water transport. And indeed there are! And not only in kidney cells but in many organs and in, who would have believed it, in bacteria as well. So now there is a multitude of porins that go around by such uninspiring names as AQP1, AQP2 and so on. But perhaps the scientifically most important finding of the research into aquaporins was the demonstration that while they do transport water, the ubiquitous H_2O, they do not transport the ubiquitous H_3O^+, the hydrated proton (Editorial 1997; Agre and Kozono 2003). (The importance of this finding will become apparent when the chemiosmotic theory of solute transport will be introduced further on.) The explanation for this discrimination is to be found in the published structure of the human aquaporin AQP1. Human aquaporin molecule is a tetramer of four identical subunits with a 20 Å long channel in each of the subunits that extends into cones at both its ends. Part of this channel is only 2.8 Å wide, this being the diameter of a single water molecule. The lining of the channel is hydrophobic with a positive charge which explains why hydronium ions are repelled from this channel. Water molecules are accommodated in the channel because of a string of carbonyl oxygens of the peptide backbone that line the channel and that keep the water molecules separated without a possibility of forming hydrogen bonds between them. Furthermore the channel is twofold symmetric in the sense that the orientation of the residues that line the channel in the distal part, that is the part that faces the interior of the cell, is opposite to the orientation of those that line the proximal part of the channel, that is the part that faces the exterior. And sure enough there are residues at this twofold axis of symmetry in the channel that cause the water molecules to flip over so that their oxygen atoms that were facing to the inside are now facing to the outside. This work has been duly accredited by conferring a Nobel Prize to its instigator, Peter Agre (Giles 2003).

Beyond the beauty and the simplicity of the aquaporin structure there is a story that is hardly possible to condense in a few paragraphs. Nor is it possible to list the pervasive medical importance of these molecules. The drama of its discovery is the fact that when first isolated in the late 1980s from red cell membranes it was not recognized as a channel forming protein for awhile. After its biological function was established it was subjected to a dramatic test with oocytes of the African aquatic frog, Xenopus laevis, that are characterized by very low water permeability. Upon introducing messenger RNA of the aquaporin into these cells and placing them in a hypoosmotic buffer, to which normal oocytes are resistant, they burst (Preston et al. 1992).

The atomic structure of aquaporin was long in coming and a structure at 3.8 Å resolution obtained by the tedious electron-diffraction of two-dimensional crystals was published under the following title: *"Structural determinants of water permeation through aquaporin-1"* (Murata et al. 2000). The aquaporin story was instrumental in extending study of porins to ones with a differing specificity such as the one responsible for transport of glycerol in E. coli. The atomic structure of this porin, GlpF, was solved by X-ray diffraction of GlpF crystals (Fu et al. 2000) and utilized in refining the human AQP1 porin (de Groot et al. 2001). Eventually crystals of AQP1 from bovine red blood cells were produced and the structure of AQP1 porin was resolved to 2.2 Å resolution. It provides the information quoted above about details of the water channel (Sui et al. 2001).

The Transport of Lactose by Escherichia coli

The fact that lactose is metabolized by E. coli, an innocuous inhabitant of human gut, was known for a long time and used in solid agar-media to distinguish colonies of this bacterium from its close relatives the Salmonellas, that can cause typhoid fever, and Shigellas, that can cause dysentery. Also the fact that microorganisms being placed in a carbon source that they cannot utilize start eventually to develop the ability to utilize it was known from research on yeast conducted in Denmark. But Gods smiled primarily on the concerted efforts to study adaptive enzyme formation in this organism by the Monod team at the Pasteur Institute, as was already mentioned in Chap. 3. And the reason, not that I presume to know Gods' minds, could well have been that their objective has been the study of the biogenesis of proteins about which hardly anything was known. What came as a complete surprise to the people involved in the study of the adaptive formation of the enzyme β-galactosidase was the fact that in defective cells, ones that did not produce an active enzyme that splits lactose, lactose was concentrated to very many times the concentration that was present in the external medium. In a sense, the scientists involved in this research got more than they bargained for. Finding a protein that facilitates selective transport of a metabolite is one thing but having to deal with the energetics of transport against a concentration gradient is another. And indeed this was left to Peter Mitchell, the father of the chemiosmotic theory of transport, and he solved it in an epoch marking work with Ian West some 20-years later (West and Mitchell 1973).In a sense the Pasteur Institute team did anticipate Mitchell's solution of their puzzle, calling the protein responsible for the transport of lactose – lac-permease, the "-ase" suffix being hitherto reserved for enzymes, considering that originally Mitchell thought of transport through membrane as a vectorial enzymatic process. Genetics apart, the most dramatic and intellectually satisfying experiment with lac-permease was the one where it was shown that round protoplasts of induced cells, that remain after the cell wall is removed by ezymatic digestion, burst when placed in a solution of lactose (Sistrom 1958).

From its humble if momentous beginnings, permease study blossomed to include transport of a multitude of hydrophilic ions and compounds in a variety of organisms (Henderson 1990). Unfortunately, semantics continues to be important to people engaged in this important if demanding field and many authentic permeases are not called permeases but transporters, secondary transporters and symporters. Thus, recently, in the same issue of a journal, two structures were reported, that are the first structures of permeases belonging to the same superfamily (Nikaido and Saier 1992; Saier 2002), one being the "Lactose Permease" and the other – "Glycerol-3-Phosphate Transporter" (Abramson et al. 2003; Huang et al. 2003). Even if one can sympathize with the authors semantic predilections, one wonders. There is an amazing structural similarity between these two permeases, considering that they transport completely different solutes. They seem to confirm a conclusion that these permeases are inserted into the membrane as monomers. This same conclusion was reached by Li and Tooth (1987) on the basis of an electron-microscopic study of filaments of lac-permease that gave a first glimpse of what a molecule of lac-permease looked-like. It is amazing how much of the secondary structure of these proteins, including the presence of a central hydrophilic region, has been predicted already in 1990 from the primary sequences of these proteins and their hydropathic profiles (Henderson 1990). The disappointment with lac-permease functioning as a monomer goes beyond the aesthetics, multimeric proteins being prettier than monomers (e.g. aquaporin, discussed above, and voltage dependent potassium channel, to be discussed below), and has, I believe, been a surprise to many. On the positive side, solving the structures of these proteins removes the pervasive distinction between the assumed mechanisms of transport of organic metabolites and of inorganic ions: the former utilising carriers and the latter pores. If anything, these two "carriers" are pores. Quite sophisticated pores, to be sure, but they have the fundamental property of a pore namely, a hydrophilic region inside a hydrophobic protein embedded in the membrane. And an ion pore need not necessarily be an unsophisticated pore either, as was shown with a chloride ion (Cl-) transporter that has some "carrier" characteristics (Mindell 2008).

Ideally, knowledge of the structure of the binding-site of lac-permease is a prerequisite to understand its specificity and its function in facilitating transport of substrates. First surprise, early on, was that the specificity of lac-permease was quite different from that of β-galactosidase and from that of lac-repressor, even if all these share a common property namely, they all bind galactosides. An expectation that these proteins will prove to be similar was disproved when their sequences were determined. What distinguishes the specificity of these proteins from lac-permease is the fact that while both bind substrates in which the terminal galactosyl is bound to the penultimate substituent in the β-configuration, irrespective of its being an ether or a thioether linkage, lac permease binds α-substituted galactosyls as well, such as the naturally occurring melibiose or a synthetic galactoside, p-NO$_2$-phenyl α-D-galactoside (Kennedy et al. 1974). As a matter of fact, the latter was used to determine the number of lac-permease molecules in a preparation. The nature of the protein, the fact that it is a membrane protein, and therefore not readily crystalline, plus the facility of producing single-point mutants in its gene, has led to the inspired and quite unique engineering feat of changing its specificity (Markgraf et al. 1985; Brooker 1990). Thus Brooker and Wilson (1985) by changing tyrosine236 to phenylalanine caused lac-permease to stop transporting lactose and to start transporting maltose instead. This was then followed by a massive effort to map the binding-site of lac-permease by means of single-point mutants. The results of this study suggested that the following two amino-acids, Glu126 and Arg144, reside in the binding-site of lac-permease (Kaback et al. 2001). And indeed the structure presented recently (Abramson et al. 2003) confirms in particular the presence of the amino-acid side-chain of Arg144 in the binding–site of lac-permease where it forms hydrogen bonds with the oxygens of hydroxyls 3 and 4 of one of the galactosyls of the substrate, this being the TDG (thiodigalactoside) where two D-galactosyl residues are β-linked by a thioether bridging the two anomeric (C1) carbons.

The Odd-Man-Out Comes to the Rescue

Mitchell was pursuing his interest in the transport of solutes through biological membranes if not single-handed, he had gifted and loyal collaborators, but in a detached manner. He was looking on the transport of solutes of all kinds as being enzymatic processes of a vectorial kind. To say that his approach was unpopular, would be an understatement. At that time all the rage about membranes was oxydative and photosynthetic phosphorylations and the schematic and highly theoretical presentations by Mitchell of the processes involved did not endear him to the biochemists brought up on biochemical cycles and embracing protein conformational changes and high-energy intermediates as an explanation of these processes. The outstanding contribution of Mitchell to the theory of oxydative phosphorylation was his insistence for the role of the membrane in these processes and of proton transport as the mediator. Fundamental in the chemiosmotic hypothesis of Mitchell is the fact that membranes are impermeable to ions, to the proton (H$^+$) in particular, and that in the membrane there exist carriers of the metabolites that he chose to call porters. Mitchell distinguishes between uniporters, symporters and antiporters. A uniporter transports only the substrate for which it is specific and is accompanied only by the concomitant transport of water that occurs elsewhere in the lipid membrane that is permeable to water. In a symporter, or antiporter, translocation of a substrate through a membrane is coupled to a transport of another compound that provides the driving force. In a symporter the compound driving the translocation is moved in the same direction and in the antiporter in an opposite direction from that of the substrate. Mitchell has singled out the proton as the all-important, if not universal, component of metabolite translocation by symporters.

Eventually Michell's chemiosmotics hypothesis was proven with lac-permease and it provides an explanation of the energetics, if not of the mechanism, of lactose accumulation by E. coli cells (Mitchell 1973). The definitive work (West and Mitchell 1973) established that the stoichiometry of lactose and hydrogen-ion transport is 1. Mitchell was not innocent of devising new names for old entities and he calls the lac-permease 'β-galactoside/H⁺ symporter'. Thus the study of lac-permease is now complicated since it tries to answer two questions at the same time namely, the mechanisms of transport of the saccharide substrate and the mechanism of transport of the proton and how are they correlated. It has been the tenet of saccharide transport by lac-permease that the substrate is not modified in the transport process. Now Naftalin et al. (2007) are advancing a model of transport of lactose by lac-permease that involves the saccharide in the symport of the proton, even if only transiently. This is a most interesting model! It was partly arrived at by performing docking experiments on computer with a molecular recognition program. If one is to accept the conclusions of this study, the saccharide binding-site of lac-permease is not a unique site but a rather flexible domain with a capacity to bind different saccharides. Be it as it may the binding site of lac-permease remains an enigma in two respects. One is that crystallographic evidence finds no ready saccharide binding-site in this protein until substrate is bound suggesting that this site is of the "induced fit" kind (Mirza et al. 2006). This study, that is an improvement on the previous one (Abramson et al. 2003), used a different crystal form of the protein and data were collected to a higher resolution. The prominence of Trp151, both in native crystal and in the crystal with the bound substrate, is suggestive that in lac-permease, similarly to other saccharide-binding proteins, van der Waals interactions are dominant. The other difficulty of this study thus far is that it does not provide evidence for the concomitant transport of the proton with the permease substrate nor for the conformational change of the pore necessary to change the exposure of the binding-site to the external solution.

Isolation of lac-permease (Fox and Kennedy 1965) antedated the work of Mitchell & West. The molecular-weight of 46,504 daltons of this protein, consisting of 417 amino-acids, was determined from the sequence of its gene (Büchel et al. 1980). Soon enough lac-permease has been expressed in plasmids and when incorporated into lipid vesicles was shown to be instrumental in both proton and substrate transport (Newman and Wilson 1980; Newman et al. 1981).

Acceptance of Mitchel's chemiosmotic hypothesis as a sine-qua-non of all transport processes was brought about by an experiment carried out by two scientists: one a famous biochemist, E. Racker, and the other the man behind the utilization of the purple membrane of Halobacterium in the study of membrane proteins, W. Stoeckenius. This work entitled "Reconstitution of Purple Membrane Vesicles Catalyzing Light-driven Proton Uptake and Adenosine Triphosphate Formation" was published as a two-page communication (Racker and Stoeckenius 1974) and was terminated with the following sentence: "They also provide a model system for energy conversion according to the chemiosmotic hypothesis." Since then there was no turning back for the chemiosmotic hypothesis (Mitchell 1973) and Mitchell was honoured by the award of the Nobel Prize in Chemistry in 1978 (Slater 2003b).

The story of Mitchell's achievements in science won't be complete without mentioning a romantic side to his personality. He would well fit into the "Enchanted Cornwall" by Daphne Du Maurier. After a stint at the Universities of Cambridge and Edinburgh, he withdrew to a farm in Bodmin, Cornwall. There he built a laboratory that became a mystery to his neighbours and a place of pilgrimage to scientist from all over the world once his chemiosmotic theory received the recognition it deserved.

Two Nobel Prizes, Forty Years Apart, to a Membrane Transporter

Working with squid axon some 50 years ago Hodgkiss and Huxley established a connection between ion conductance and electrical potential in neurons. They found that transport of ions through the membrane is extremely selective to wit flow of sodium into the cell makes the voltage more positive while the eflux of potassium from the cell makes it more negative. They assumed, furthermore, that flow of sodium and potassium and the concomitant change in the electrical potential is mediated by different molecules or systems. For this outstanding insight into the mechanism of the creation of the action potential they were awarded the Nobel Prize in Physiology and Medicine in 1963. Electrophysiology is an extremely specialized science, even more then than now, and anybody not versed in the arcana of this discipline will have a hard time to understand their respective Nobel Prize Lectures not to mention the original papers. Forty years later Roderick MacKinnon, an electrophysiologist who has switched horses in the middle of his career and became an X-ray crystallographer, has solved the structure of one such protein namely, the one responsible for the transport of potassium, and was honoured with a Nobel Prize in Chemistry that he shared with Peter Agre of the aquaporin fame (Giles 2003).

And oh, what a wonderful protein is this Voltage Dependent potassium Channel! Science progresses when scientists need to revise their concepts. In the realm of membrane transport two mechanisms of solute transport were considered: passage through a solute specific pore or the solute being piggy-back carried by a carrier. Now, Voltage Dependent potassium Channel is a transporter that uses both mechanisms: the membrane crossing potassium specific pore that is created by the association of four identical monomers to form a tetramer that is the functional molecule and four "paddles", that is hydrophobic protrusions from each of the monomers that form the molecule, that are positioned so that they can extend to both the internal and the external face of the membrane to carry the charge and in doing so form or collapse the electrical potential. Nobody, but nobody, has imagined that this is how the charge is transported and nobody, but nobody, has ever suggested such a mechanism for a transporter (Sigworth 2003; Jiang et al. 2003a, b).

If I choose to terminate this discourse on binding-sites by describing the structure of the voltage-gated potassium channel, it is for two reasons. Beyond the beauty of the structure of the potassium channel, which would be reason enough to try to describe it, such channels share one property with binding-sites – specificity. How is it that this particular pore transports a potassium ion but does not transport the smaller sodium ion.

Voltage dependent potassium channel similarly to aquaporin is also a tetramer of a fourfold symmetry. But while the four subunits of aquaporin each form a pore, in the voltage dependent potassium channel only one pore is on the fourfold axis between the four subunits. This potassium channel is a much more complicated and, may one say, sophisticated channel than the water channel of aquaporin. MacKinnon has solved its structure having brought to it an unimaginable number of laboratory approaches to bear. Thus already in 1991 he published a paper with self explanatory title "Determination of the subunit stoichiometry of a voltage-activated potassium channel" (MacKinnon 1991) in which the stoichiometry was determined with the help of a scorpion toxin and found to be the magic number of 4. This then established that the protein forming the channel was a tetramer.

As was mentioned above electrophysiology is a very specialized science and as such it employs a nomenclature not readily available to simple mortals. Thus, I would advice the unprepared reader to read beforehand a short review in FEBS Letters by none other than the same Roderick MacKinnon (2003). To summarize, it deals with three aspects of the seminal papers (Jiang et al. 2003a, b): ion conduction, gating of the pore and voltage sensing. It has to be realized that this channel transport 10^8 potassium ions per second and that it favors the potassium ion over the sodium ion by a factor of at least 10^4. To quote MacKinnon "All known K^+ channels are related members of a single protein family. They are found in bacterial, archeal, and eukaryotic cells – both plant and animal – and their amino acid sequences are very easy to recognize because potassium channels contain a highly conserved sequence called the K^+ channel signature sequence. This sequence forms a structural element known as the selectivity filter, which prevents the passage of Na^+ ions but allows K^+ ions to conduct across the membrane at rates approaching the diffusion limit" (Fig. 5.1). Now, gating of the pore involves a conformational change that opens the pore. Gating by voltage is one of the many different gating agents of potassium channels and gating can be caused as well by the binding of another specific ion or of an organic molecule. As for the voltage gating part of the evolving story, this is the cherry in the pie and I cannot hope to improve on what Sigworth (2003) had to say about it in his comments entitled "Life Transistors" in the same issue of Nature that published the structure of this marvelous protein. Thus I quote, "The molecular structures within ion channels that sense the membrane voltage have remained obscure for the 50 years since Hodgkin and Huxley first described their function. But the voltage sensors have at last been made visible, in the X-ray structure of a potassium ion channel.......As voltage sensing devises, these channels can perform better than their electronic counterparts …….. Their high sensitivity to voltage is important, because cellular voltage changes are small. ……. Determination of the structure of a voltage-gated channel has been long in coming. The five-year effort in the MacKinnon laboratory involved trials of many channel proteins, none of which formed either two-dimensional or three-dimensional crystals. ………. But the structure of KvAP's voltage sensor, so simple and, with hindsight, so obvious, is a wonderful end to a 50-year-old mystery."

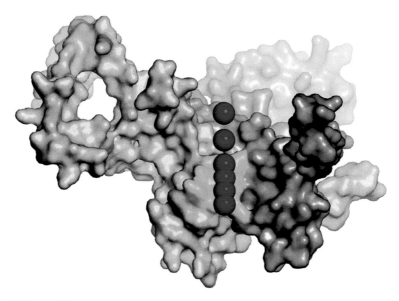

Fig. 5.1 Space-filling model of the Selectivity Filter of the Voltage Dependent Potassium Channel (The model was constructed with the crystallographic coordinates 1D 3LMN in the Protein Data Bank)

Fig. 5.2 The tetragonal structure of the Voltage Dependent Potassium Channel (Figure and legend are reproduced with permission from Jiang et al. 2003a, b)

For once I shall depart from a principle I adopted in choosing figures to illuminate this text namely, that they be comprehensible to the reader and that therefore figures that employ graphic presentations comprehensible to specialists only are excluded. The exception is the figure on the first page of the article by Jiang et al. (2003a, b) where the tetragonal structure of the voltage-dependent K+ channel is apparent even to an untrained eye (Fig. 5.2).

References

Abramson J, Smirnova I, Kasho V, Verbner G, Kaback HR, Iwata S (2003) Structure and mechanism of the lactose permease of Escherichia coli. Science 301:610–615

Agre P, Kozono D (2003) Aquaporin water channels: molecular mechanisms for human diseases. FEBS Lett 555:72–78

Brooker RJ (1990) The lactose permease of Escherichia col. Res Microbiol 249:309–315

Brooker RJ, Wilson TH (1985) Isolation and nucleotide sequencing of lactose carrier mutants that transport maltose. Proc Natl Acad Sci U S A 82:3959–3963

Büchel DE, Groneborn B, Müller-Hill B (1980) Sequence of the lactose permease gene. Nature 283:541–545

Davson H, Danielli JF (1943) The permeability of natural membranes. University Press, Cambridge

de Groot BL, Engel A, Grubmüller H (2001) A refined structure of human aquaporin-1. FEBS Lett 504:206–211

Editorial (1997) A biological water filter. Nat Struct Biol 4:245–246

Fox CF, Kennedy EP (1965) Specific labelling and partial purification of the M protein, a component of the b-galactoside transport system of Escherichia coli. Biochemistry 54:891–899

Fu D, Libson A, Mierke IJ, Weitzman C, Nollert P, Krucinski J, Stroud RM (2000) Science 290:481–486

Giles J (2003) Channel hoppers land chemistry Nobel. Nature 425:651

Henderson PJF (1990) Proton-linked sugar transport systems in bacteria. J Bioenerg Biomembr 22:525–569

Hodgkin AL, Huxley AF (1952) A quantitative description of membrane current and its application to conduction and excitation in nerve. J Physiol 117:500–544

Huang Y, Lemieux MJ, Song J, Auer M, Wang D-N (2003) Structure and mechanism of the glycerol-3-phosphate transporter from Escherichia coli. Science 301:616–620

Jiang Y, Lee A, Chen J, Ruta V, Cadene M, Chait BT, Mackinnon R et al (2003a) X-ray structure of a voltage-dependent K+ channel. Nature 423:33–41

Jiang Y, Ruta V, Chen J, Lee A, Mackinnon R (2003b) The principle of gating charge movement in a voltage dependent K+ channel. Nature 423:42–48

Kaback HR, Sahin-Toth M, Weinglass AB (2001) The Kamikaze approach to membrane transport. Nat Rev Mol Cell Biol 2:610–622

Keilin D (1925) On cytochrome, a respiratory pigment, common to animals, yeast, and higher plants. Proc R Soc B Biol Sci 98:312–339

Kennedy EP, Rumley MK, Armstrong JB (1974) Direct measurement of the binding of labelled sugars to the lactose permease M protein. J Biol Chem 249:33–37

Li J, Tooth P (1987) Size and shape of the Escherichia coli lactose permease measured in filamentous arrays. Biochemistry 26:4816–4826

MacKinnon R (1991) Determination of the subunit stoichiometry of a voltage- activated potassium channel. Nature 350:232–235

MacKinnon R (2003) Potassium channels. FEBS Lett 555:62–65

Markgraf M, Bocklage H, Müller-Hill B (1985) A change of threonine266 to isoleucine in the lac permease of Escherichia coli diminishes the transport of lactose and increases the transport of maltose. Mol Gen Genet 198:473–475

Mindell JA (2008) The chloride channel's appendix. Nat Struct Mol Biol 15:781–783

Mirza O, Guan L, Verner G, Iwata S, Kaback HR (2006) Structural evidence for induced fit and a mechanism for sugar/H$^+$ symport in LacY. EMBO J 25:1177–1183

Mitchell P (1973) The chemiosmotic theory of transport and metabolism. In: Mechanisms in Bioenergetics. Academic, New York, pp 177–201

Murata K, Mitsuoka K, Hirai T, Walz T, Agre P, Heymann JB, Engel A (2000) Structural determinants of water permeation through aquaporin-1. Nature 407:599–605

Naftalin RJ, Green N, Cunningham P (2007) Lactose permease H$^+$- lactose: symporter mechanical switch or Brownian Ratchet ? Biophys J 92:3474–3491

Newman MJ, Wilson TH (1980) Solubilization and reconstitution of the lactose transport system from Escherichia coli. J Biol Chem 255:10583–10586

Newman MJ, Foster DL, Wilson TH, Kaback HR (1981) Purification and reconstitution of functional lactose carrier from Escherichia coli. J Biol Chem 256:11804–11808

Nikaido H, Saier MHJ (1992) Transport proteins in bacteria: common themes in their design. Science 258:936–942

Preston GM, Carroll TP, Guggino WB, Agre P (1992) Science 256:385–387

Racker E, Stoeckenius W (1974) Reconstitution of purple membrane vesicles catalyzing light-driven proton uptake and adenosine triphosphate formation. J Biol Chem 249:662–663

Rickenberg HV, Cohen GN, Buttin G, Monod J (1956) Ann Inst Pasteur 91:829

Saier MHJ (2002) Families of transporters and their classification. In: Transmembrane transporters. Wiley Liss, New York, pp 1–17

Sigworth FJ (2003) Structural biology: life's transistors. Nature 423:21–22

Sistrom WR (1958) On the physical state of the intracellularly accumulated substrates of β-galactoside-permease in Escherichia coli. Biochim Biophys Acta 29:579–587

Slater EC (2003a) Keilin, cytochrome, and the respiratory chain. J Biol Chem 278:16455–16461

Slater EC (2003b) Metabolic gardening. Nature 422:816–817

Sui H, Han B-G, Lee JK, Walian P, Jap BK (2001) Structural basis of water-specific transport through the AQP1 water channel. Nature 414:872–877

West IC, Mitchell P (1973) Stoichiometry of lactose-H$^+$ symport across the plasma membrane of Escherichia coli. Biochem J 132:587–592

Membrane Proteins – Epilogue

It is ironic that Mitchell's hypothesis has been proven in lac-permease and purple membrane vesicles, considering that both he and the biochemists have been concerned with the mechanism of oxidation and the synthesis of ATP in the mitochondria. It is doubly-ironic that in a recent article (Glancy et al. 2015) it is stated that the proton-motive force is expressed as a trans-membrane electrical potential on the network of mitochondria and that this is what drives muscle contraction. In this context a review by Skulatchev (Skulachev 2001), an old-hand at oxidative-phosphorylation and electrical potential is highly recommended.

A landmark in connecting proton transport to ATP synthesis in the mitochondria has been the solution of the structure of the F_1-ATPase of mitochondria (Abrahams et al. 1994). This is a huge membrane-spanning protein composed of 9 subunits. The devise by which this huge protein synthesizes ATP from ADP and inorganic phosphate is best described as a carrousel and the authors rightly call the mechanism by which it is realized rotational catalysis. ATP synthesis utilizes protons. As a matter of fact, to quote Abrahams et al., "About three protons flow through the membrane per ATP synthesized". In mitochondria the protons are supplied by cytochrome oxidase another huge membrane protein and, as mentioned already in the introductory section of this chapter, part of the respiratory chain (Calhoun et al. 1994).

© Springer International Publishing Switzerland 2016
J. Yariv, *The Discreet Charm of Protein Binding Sites*,
DOI 10.1007/978-3-319-24996-4

References

Abrahams JP, Leslie AGW, Lutter R, Walker JE (1994) Structure at 2.8 Å resolution of F_1-ATPase from bovine heart mitochondria. Nature 370:621–628

Calhoun MW, Thomas JW, Gennis RB (1994) The cytochrome oxidase superfamily of redox-driven proton pumps. Trends Biochem Sci 19:325–330

Glancy B, Hartnell LM, Malide D, Yu Z-X, Combs CA, Connelly PS, Subramaniam S, Balaban ES (2015) Nature 532:617–620

Skulachev VP (2001) Mitochondrial filaments and clusters as intracellular power-transmitting cables. Trends Biochem Sci 26:23–29